ELEMENTARY ELECTROMAGNETIC THEORY

VOLUME 3

MAXWELL'S EQUATIONS AND THEIR CONSEQUENCES

BOOKS BY DR. CHIRGWIN, PROFESSOR KILMISTER AND DR. PLUMPTON ON ELECTROMAGNETIC THEORY

ELEMENTARY ELECTROMAGNETIC THEORY

Volume 1 STEADY ELECTRIC FIELDS AND CURRENTS

Volume 2 MAGNETIC FIELDS, SPECIAL RELATIVITY AND POTENTIAL THEORY

Volume 3 MAXWELL'S EQUATIONS AND THEIR CONSEQUENCES

ELEMENTARY ELECTROMAGNETIC THEORY

IN THREE VOLUMES

VOLUME 3
MAXWELL'S EQUATIONS AND THEIR CONSEQUENCES

B. H. CHIRGWIN
Queen Mary College, London

C. PLUMPTON
Queen Mary College, London

AND

C. W. KILMISTER
King's College, London

PERGAMON PRESS
OXFORD · NEW YORK · TORONTO
SYDNEY · BRAUNSCHWEIG

Pergamon Press Ltd., Headington Hill Hall, Oxford
Pergamon Press Inc., Maxwell House, Fairview Park, Elmsford, New York 10523
Pergamon of Canada Ltd., 207 Queen's Quay West, Toronto 1
Pergamon Press (Aust.) Pty. Ltd., 19a Boundary Street, Rushcutters Bay, N.S.W. 2011, Australia
Vieweg & Sohn GmbH, Burgplatz 1, Braunschweig

First edition 1973

Library of Congress Cataloging in Publication Data

Chirgwin, Brian H.
Elementary electromagnetic theory.
CONTENTS: v. 1. Steady electric fields and currents.
v. 2. Magnetic fields, special relativity and potential theory.
v. 3. Maxwell's equations and their consequences.
1. Electromagnetic theory. 2. Potential, Theory of. I. Plumpton, Charles, joint author.
II. Kilmister, Clive William, joint author.
III. Title.
QC670.C4 530.1'41 70-129631
ISBN 0 08 016079 4 (v. 1)

Printed in Hungary

ISBN 0 08 017120 6 (Hard cover)
ISBN 0 08 017121 4 (Flexicover)

CONTENTS

PREFACE TO VOLUME 3

THIS is the third of three volumes intended to cover the electromagnetism and potential theory usually included in an undergraduate's course of study. These books are intended only as an introduction to electromagnetism and have been prompted by discussion with first-, second- and third-year undergraduates.

The general scheme of the volumes is to start with the simple case of steady fields and to develop the appropriate generalizations when this constraint is relaxed. Thus, in the first volume we started with the fields associated with stationary charges and relaxed the stationary condition to allow consideration of the flow of steady currents in closed circuits.

In the first volume we considered the experimental results which require mathematical explanation and discussion, in particular those referring to phenomena which suggest that the simple Newtonian concepts of space and time are not fully valid. Then we considered steady state fields and dealt next with electrostatics including dielectrics, energy theorems, uniqueness theorems, and ended that volume with a chapter on the steady flow of electric currents. SI units were used throughout, although the older systems were briefly mentioned.

In the second volume we first of all considered the magnetic field of steady currents—magnetostatics. This was followed by a chapter on the methods of solving potential problems drawn from electrostatics, magnetism, current flow and gravitation. Relaxing the constraint of stationary steady currents, we were led to consider electromagnetic induction when the current strengths in closed circuits vary or when the circuits move. This led to the necessity of considering the breakdown of Newtonian ideas and the introduction of special relativity. When we further relaxed the contraint of closed circuits and considered the motion of charges we were led to introduce the displacement current because of the relativistic theory already set up, and so were led to Maxwell's equations.

In this third volume we consider the implications of Maxwell's equations, such as electromagnetic radiation in simple cases, and deal further with the relation between Maxwell's equations and the Lorentz transformation.

We assume that the readers are conversant with the basic ideas of vector analysis, including vector integral theorems.

Our thanks are due to the University of Oxford, to the Syndics of the Cambridge University Press and to the Senate of the University of London for permission to use questions set in their various examinations.

CHAPTER 11

ELECTROMAGNETIC WAVES

11.1 Plane electromagnetic waves

We have shown earlier, pp. 419–420 of Volume 2, that the various field vectors and potentials of Maxwell's theory each satisfy a wave equation. The general solution of the wave equation is of little practical value; particular solutions are much more important and contribute a great deal to our understanding of certain physical phenomena. One simple solution of Maxwell's equations corresponds to a *plane wave;* its importance stems from the fact that complicated fields can be constructed from plane wave solutions by application of Fourier's theorem. Further, a plane wave can be produced experimentally to a fair degree of accuracy in the shape of a parallel beam of light.

For a plane-wave solution we seek a solution of Maxwell's equations in which the field quantities depend upon distance along the direction of propagation of the beam, and upon the time, but are uniform in directions across the beam. We choose the direction of the frame of reference so that the x-axis gives the direction of propagation of the beam, and consider only free space in which there is no charge or current density. With these restrictions Maxwell's equations reduce to eight component equations as follows, and all quantities depend, in general, only on x and t.

$$\operatorname{div} \boldsymbol{D} = \varrho = 0; \qquad \frac{\partial D_x}{\partial x} = \varepsilon_0 \frac{\partial E_x}{\partial x} = 0. \tag{11.1}$$

$$\operatorname{div} \boldsymbol{B} = 0; \qquad \frac{\partial B_x}{\partial x} = 0. \tag{11.2}$$

$$\mathbf{curl}\, \boldsymbol{E} + \frac{\partial \boldsymbol{B}}{\partial t} = \boldsymbol{0}; \qquad \frac{\partial B_x}{\partial t} = 0, \tag{11.3}$$

$$-\frac{\partial E_z}{\partial x} + \frac{\partial B_y}{\partial t} = 0, \tag{11.4}$$

$$\frac{\partial E_y}{\partial x}+\frac{\partial B_z}{\partial t} = 0. \tag{11.5}$$

$$\mathbf{curl}\ \boldsymbol{H}-\frac{\partial \boldsymbol{D}}{\partial t} = \boldsymbol{0}; \qquad \varepsilon_0\frac{\partial E_x}{\partial t} = 0, \tag{11.6}$$

$$-\frac{1}{\mu_0}\frac{\partial B_z}{\partial x}-\varepsilon_0\frac{\partial E_y}{\partial t} = 0, \tag{11.7}$$

$$\frac{1}{\mu_0}\frac{\partial B_y}{\partial x}-\varepsilon_0\frac{\partial E_z}{\partial t} = 0. \tag{11.8}$$

Equations (11.1) and (11.6), (11.2) and (11.3) show that both E_x, B_x are independent of x and of t. This means that they must be constants. Since we are primarily interested in varying fields we take these constants to be zero. The remaining components B_y, B_z, E_y, E_z all satisfy the wave equation

$$\frac{\partial^2 f}{\partial x^2} = \frac{1}{c^2}\frac{\partial^2 f}{\partial t^2},$$

where $\varepsilon_0\mu_0 = 1/c^2$. The general solution of this equation is

$$f(x, t) = F(x-ct)+G(x+ct)$$

where F and G are arbitrary functions of a single argument. The term $F(x-ct)$ corresponds to a wave travelling in the positive x-direction with speed c, and $G(x+ct)$ to another wave travelling in the negative x-direction also with speed c. In our discussion of the propagation of a wave we consider, without loss of generality, only one of these terms.

We see that eqns. (11.4), (11.8) relate E_z, B_y and that eqns. (11.5), (11.7) relate E_y, B_z, one pair being independent of the other pair. Hence we write for $\boldsymbol{B}$,

$$B_x = 0, \quad B_y = F_1(x-ct), \quad B_z = F_2(x-ct); \tag{11.9}$$

and obtain for $\boldsymbol{E}$,

$$E_x = 0, \quad E_y = cF_2(x-ct), \quad E_z = -cF_1(x-ct). \tag{11.10}$$

These results show

(i) that both fields are perpendicular to the direction of propagation, i.e. the waves are *transverse*; and

(ii) that $\boldsymbol{i}\times\boldsymbol{E} = c\boldsymbol{B}$. This latter relation shows that $\boldsymbol{E}$, $\boldsymbol{B}$ and the direction of propagation, $\boldsymbol{i}$, form a right-handed set of mutually perpendicular directions.

The most important case of the plane-wave solution is the *harmonic plane wave*. This is one in which the time variation of all the field quantities is a simple harmonic variation with frequency $\omega/(2\pi)$ and corresponds to

$$F_1(x-ct) = a\cos(\omega t-\omega x/c+\alpha), \tag{11.11}$$

where α is a phase angle (constant). The wavelength λ is given by $\lambda = 2\pi c/\omega$, and the wave number is $k = \omega/c = 2\pi/\lambda$. The most convenient method of handling these harmonically varying quantities is to use complex exponentials (see § 9.7, Vol. 2) so that the time variation is given by the factor $e^{i\omega t}$, which can in most cases be removed by cancellation on both sides of an equation. Using this method we take

$$B_y = Ae^{i(\omega t-kx)}, \qquad B_z = Be^{i(\omega t-kx)} \tag{11.12}$$

so that

$$E_y = cBe^{i(\omega t-kx)}, \quad E_z = -cAe^{i(\omega t-kx)}, \tag{11.13}$$

where A, B are (constant) complex numbers whose arguments give the phase constant α, and whose moduli give the amplitude of the oscillations. The basis of this method is that physical values are given by the real parts of the complex expressions (sometimes by the imaginary parts); its main advantage lies in the fact that differentiation w.r. to the time $(\partial/\partial t)$ can be replaced by multiplication by $i\omega$, and differentiation with respect to the space coordinate $(\partial/\partial x)$ is replaced by multiplication by $-ik$.

The solution given by (11.12), (11.13) corresponds to (mutually orthogonal) vectors $\boldsymbol{E}$, $\boldsymbol{B}$ at any point which change their magnitude and direction with time. If we imagine the vector $\boldsymbol{B}$ (or $\boldsymbol{E}$) as a directed segment of a line drawn from the point, then (11.12) and (11.13) imply that the end of the vector traces out an ellipse. This wave is therefore said to be *elliptically polarized*.

Since we are considering physical values the real parts are, for B_y and B_z,

$$B_y = a\cos(\omega t-kx+\alpha), \quad B_z = b\cos(\omega t-kx+\beta).$$

We solve these two equations for $\cos(\omega t-kx)$ and $\sin(\omega t-kx)$ and obtain

$$\frac{\cos(\omega t-kx)}{aB_z\sin\alpha-bB_y\sin\beta} = \frac{\sin(\omega t-kx)}{aB_z\cos\alpha-bB_y\cos\beta} = \frac{1}{ab\sin(\alpha-\beta)}.$$

Hence

$$(aB_z\sin\alpha-bB_y\sin\beta)^2+(aB_z\cos\alpha-bB_y\cos\beta)^2 = a^2b^2\sin^2(\alpha-\beta),$$

i.e. $$b^2B_y^2-2abB_yB_z\cos(\alpha-\beta)+a^2B_z^2 = a^2b^2\sin^2(\alpha-\beta).$$

This relation implies that the end of the vector $\boldsymbol{B}$ traces out an ellipse (see Fig. 11.1a).

(In all these cases illustrated in Fig. 11.1 the direction of propagation of the wave is towards the reader.) The special cases are illustrated in the figure as follows:

Figure 11.1 (b) (i). In this case $a = b$, $\beta = \alpha + \pi/2$. The ellipse becomes a circle which is traced out as indicated. Such a wave is *circularly polarized* in the right-hand sense.

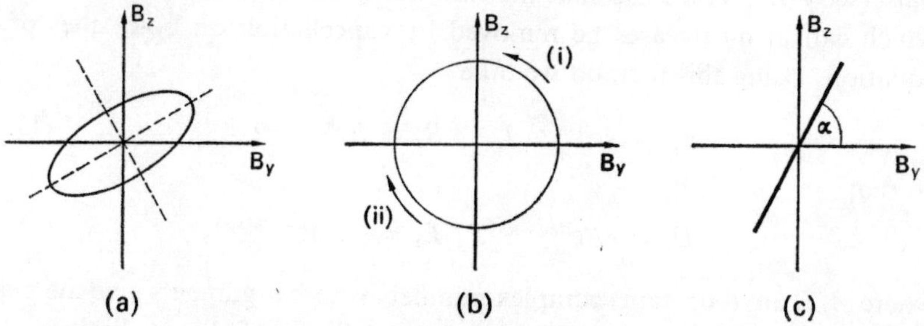

FIG. 11.1

Figure 11.1 (b) (ii). In this case $a = b$, $\beta = \alpha - \pi/2$. The ellipse is a circle traced out in the opposite sense; this wave has left-hand *circular polarization*.

Figure 11.1 (c). In this case the arguments of both A and B are equal and the ellipse degenerates to a straight line. This wave is a *plane-polarized* wave, the orientation of the plane being given by the angle α.

11.2 Reflection and transmission: normal incidence

Light is reflected from the smooth surface of a sheet of metal; at the surface of glass, water, or other transparent media, some of the incident light is reflected and some is transmitted—after refraction at the surface—if the light is incident obliquely on the surface. These general phenomena are familiar in daily experience and the laws governing the directions of the rays are well known in elementary physics:

$$\text{reflection:}\quad \theta' = \theta;$$

$$\text{refraction:}\quad n \sin \theta'' = \sin \theta \qquad \text{(Snell's law),}$$

where n is the refractive index of the (optically dense) medium occupying the region below the surface as illustrated in Fig. 11.2. Also, all three rays

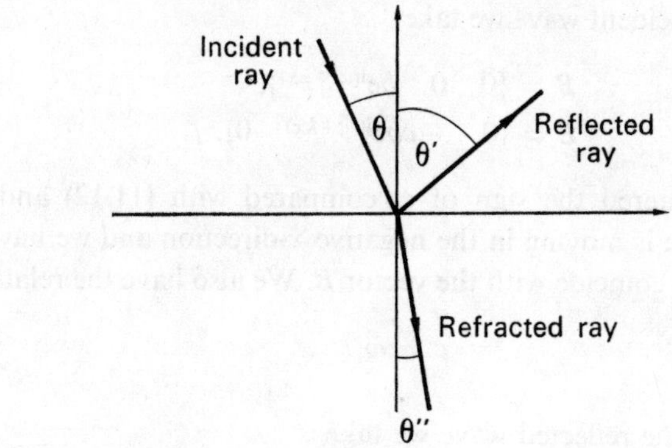

FIG. 11.2

and the normal lie in one plane—the *plane of incidence*. The special case we consider in this section is that of normal incidence in which $\theta = \theta' = \theta'' = 0$.

We give an account of these phenomena in terms of Maxwell's theory and the conditions which must apply at the surface of discontinuity separating the two regions. We assume that the permittivity of the upper region $(x > 0)$ is ε_0 (i.e. the medium is a vacuum or a gas such as air) and that the permittivity of the lower region $(x < 0)$ is $K\varepsilon_0$, where K is the dielectric constant of the optically denser medium. We take the magnetic permeabilities to be μ_0 for both $x > 0$ and $x < 0$, and assume that there is no charge or current on the surface of separation, Fig. 11.3.

To discuss the process of reflection and refraction we assume an expression obtained from eqns. (11.12) and (11.13) for a plane harmonic wave, and take one such expression for each of the incident, reflected and refracted

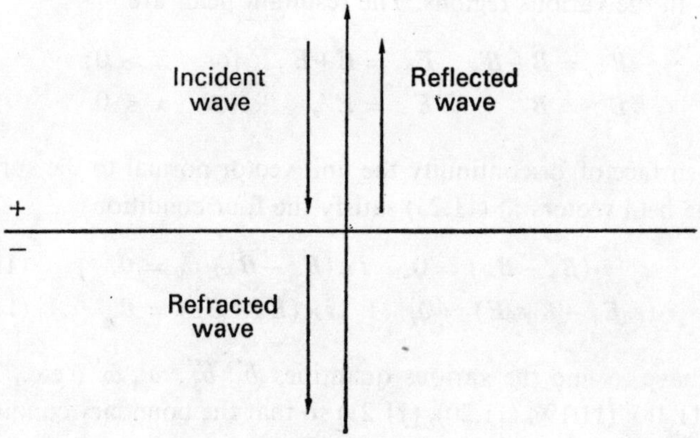

FIG. 11.3

waves. For the incident wave we take

$$\boldsymbol{B} = \{0 \quad 0 \quad b\mathrm{e}^{\mathrm{i}(\omega t+kx)}\}, \tag{11.14}$$
$$\boldsymbol{E} = \{0 \quad -cb\mathrm{e}^{\mathrm{i}(\omega t+kx)} \quad 0\}. \tag{11.15}$$

Here we have altered the sign of *c*, compared with (11.12) and (11.13), because this wave is moving in the negative *x*-direction and we have chosen the *z*-direction to coincide with the vector **B**. We also have the relation

$$c = \omega/k \tag{11.16}$$

for the velocity.

Similarly, for the reflected wave we take

$$\boldsymbol{B}' = \{0 \quad b_2' \quad b_3'\}\,\mathrm{e}^{\mathrm{i}(\omega' t-k'x)}, \tag{11.17}$$
$$\boldsymbol{E}' = \{0 \quad cb_3' \quad -cb_2'\}\,\mathrm{e}^{\mathrm{i}(\omega' t-k'x)}. \tag{11.18}$$

Here we must have

$$c = \omega'/k'. \tag{11.19}$$

For the refracted wave we take

$$\boldsymbol{B}'' = \{0 \quad b_2'' \quad b_3''\}\,\mathrm{e}^{\mathrm{i}(\omega'' t+k''x)}, \tag{11.20}$$
$$\boldsymbol{E}'' = \{0 \quad -vb_3'' \quad vb_2''\}\,\mathrm{e}^{\mathrm{i}(\omega'' t+k''x)}. \tag{11.21}$$

Here v is the velocity of propagation of the wave in the optically dense medium given by

$$\mu_0\varepsilon_0 K = 1/v^2, \quad v = \omega''/k'',$$

i.e.

$$v = \omega''/k'' = c/\sqrt{K}. \tag{11.22}$$

We know from the previous section that these expressions satisfy Maxwell's equations in the various regions. The resultant fields are

$$\begin{aligned} \boldsymbol{B}_+ &= \boldsymbol{B}+\boldsymbol{B}', \quad \boldsymbol{E}_+ = \boldsymbol{E}+\boldsymbol{E}', \quad \text{for} \quad x > 0; \\ \boldsymbol{B}_- &= \boldsymbol{B}'', \quad \boldsymbol{E}_- = \boldsymbol{E}'', \quad \text{for} \quad x < 0. \end{aligned} \tag{11.23}$$

At the surface of discontinuity the unit vector normal to the surface is **i**, and so the field vectors in (11.23) satisfy the four conditions

$$\boldsymbol{i}\cdot(\boldsymbol{B}_+-\boldsymbol{B}_-) = 0, \quad \boldsymbol{i}\times(\boldsymbol{B}_+-\boldsymbol{B}_-)/\mu_0 = \boldsymbol{0}, \tag{11.24, 25}$$
$$\boldsymbol{i}\cdot(\varepsilon_0\boldsymbol{E}_+-K_0\varepsilon_0\boldsymbol{E}) = 0, \quad \boldsymbol{i}\times(\boldsymbol{E}_+-\boldsymbol{E}_-) = \boldsymbol{0}. \tag{11.26, 27}$$

We now have to find the various quantities b_2', b_3'', ω', ω'', etc., in eqns. (11.17), (11.18), (11.19), (11.20), (11.21) so that the boundary conditions are satisfied at all times.

First we notice that, since $x = 0$ on the boundary, the field vectors are constant there except for the time factors $e^{i\omega t}$, $e^{i\omega' t}$, $e^{i\omega'' t}$. Because the boundary conditions must apply for all, arbitrary, values of t, the time factors must be identical, so that

$$\omega = \omega' = \omega''. \tag{11.28}$$

We deduce from this that incidence on the surface does not alter the frequency of the wave.

We now substitute the vectors into (11.24)–(11.27) and obtain the following results.

(11.24): $0 = 0.$

(11.25): three component equations, $0 = 0,$

$$-b - b_3' + b_3'' = 0, \tag{11.29}$$

$$b_2' - b_2'' = 0. \tag{11.30}$$

(11.26): $0 = 0,$

(11.27): three component equations $0 = 0$

$$cb_2' - vb_2'' = 0. \tag{11.31}$$

$$-cb + cb_3' + vb_3'' = 0. \tag{11.32}$$

These give the results

$$b_2' = b_2'' = 0,$$

$$b_3' = \frac{c-v}{c+v}\, b, \qquad b_3'' = \frac{2c}{v+c}\, b. \tag{11.33}$$

This last result can be expressed in alternative forms

$$\left.\begin{aligned} b_3' &= \frac{\sqrt{K}-1}{\sqrt{K}+1}\, b = \frac{n-1}{n+1}\, b, \\ b_3'' &= \frac{2\sqrt{K}}{\sqrt{K}+1}\, b = \frac{2n}{n+1}\, b, \end{aligned}\right\} \tag{11.34}$$

(where we have used the result $n^2 = K$, which we derive later, for the refractive index). The expressions in (11.34) give the amplitudes of the reflected and refracted waves in terms of the amplitude of the incident wave.

The Poynting vector $\boldsymbol{E} \times \boldsymbol{H}$ gives the density of energy flow corresponding to a given field; when we use the complex notation we can obtain the mean value of this flow as the real part of the expression $\frac{1}{2}\boldsymbol{E} \times \boldsymbol{H}^*$ (see Vol. 2, p. 406) and evaluate the energy carried by the incident, reflected and refracted rays. (Here the star * denotes complex conjugate.) The densities are

$$\begin{aligned} &\text{incident} \quad cb^2/(2\mu_0) = I, \\ &\text{reflected} \quad cb_3'^2/(2\mu_0) = \frac{(n-1)^2}{(n+1)^2} I, \\ &\text{refracted} \quad vb_3''^2/(2\mu_0) = \frac{1}{n}\,\frac{4n^2}{(n+1)^2} I \end{aligned} \tag{11.35}$$

in the appropriate direction along the x-axis. The mean energy flow on the positive side of the surface of separation ($x > 0$) is $\mathrm{Re}\,(\frac{1}{2}\boldsymbol{E}_+ \times \boldsymbol{H}_+^*)$, where

$$\frac{1}{2}\boldsymbol{E}_+ \times \boldsymbol{H}_+^* = \frac{1}{2\mu_0}\boldsymbol{E}_+ \times \boldsymbol{B}_+^* = \frac{1}{2\mu_0}(\boldsymbol{E}+\boldsymbol{E}')\times(\boldsymbol{B}^*+\boldsymbol{B}'^*).$$

We substitute the expressions (11.15–18) for the field vectors, remembering that $b_2' = 0$, and obtain

$$\frac{1}{2}\boldsymbol{E}_+ \times \boldsymbol{H}_+^* = \frac{c}{2\mu_0}(-b\,\mathrm{e}^{ikx} + b_3'\,\mathrm{e}^{-ikx})\,(b\,\mathrm{e}^{-ikx} + b_3'^*\,\mathrm{e}^{ikx})\,\{1 \quad 0 \quad 0\}.$$

The coefficient of the unit vector $\{1 \quad 0 \quad 0\}$ is

$$\frac{c}{2\mu_0}\{-b^2 + b_3'b_3'^* + b(b_3'\,\mathrm{e}^{-2ikx} - b_3'^*\,\mathrm{e}^{2ikx})\}. \tag{11.35a}$$

To obtain the energy flow we need the real part of this, which is (taking b to be real)

$$\frac{c}{2\mu_0}(-b^2 + b_3'b_3'^*).$$

Hence the energy transported toward the surface is that of the incident ray, from the term $-b^2$, and away from the surface is that of the reflected ray. The remainder of (11.35a) is imaginary and makes no contribution to the energy flow. Since $\boldsymbol{E}_- = \boldsymbol{E}''$, $\boldsymbol{B}_- = \boldsymbol{B}''$ for $x < 0$, the energy transported away from the surface into the medium is that in the refracted ray. We see from eqn. (11.35) that the sum of the reflected and refracted energies is equal to the incident energy. A *reflection coefficient* R and a *transmission coefficient* T can be defined as the ratio of the corresponding energy to the incident energy and have the values

$$R = \frac{(n-1)^2}{(n+1)^2}, \quad T = \frac{4n}{(n+1)^2}.$$

Example 1. The field vectors of a train of plane-polarized electromagnetic waves travelling *in vacuo* are given at the point (x, y, z) by

$$\boldsymbol{E} = \boldsymbol{E}_0 \cos \varkappa(z-ct), \quad \boldsymbol{B} = \boldsymbol{B}_0 \cos \varkappa(z-ct).$$

Prove that

$$\boldsymbol{k} \times \boldsymbol{E}_0 = c\boldsymbol{B}_0, \quad c(\boldsymbol{B} \times \boldsymbol{k}) = \boldsymbol{E}_0.$$

If such a train of waves is incident normally on a perfectly conducting plane $z = 0$, show that the density of the induced surface current is

$$(2/\mu_0)(\boldsymbol{B}_0 \times \boldsymbol{k}) \cos \varkappa ct$$

and prove that the average pressure on the plane is $2B_0^2/\mu_0$.

First we find the relations satisfied by $\boldsymbol{E}_0$, $\boldsymbol{B}_0$ in order that the expressions given satisfy Maxwell's equations. We consider the equations in succession:

$$\text{div } \boldsymbol{E} = 0; \quad \text{div } \boldsymbol{E} = \boldsymbol{E}_0 \cdot \textbf{grad} \cos \varkappa(z-ct) = -(\boldsymbol{E}_0 \cdot \boldsymbol{k})\varkappa \sin \varkappa(z-ct) = 0.$$

Therefore

$$\boldsymbol{E}_0 \cdot \boldsymbol{k} = 0, \tag{1}$$

Similarly the equation div $\boldsymbol{B} = 0$ implies that

$$\boldsymbol{B}_0 \cdot \boldsymbol{k} = 0. \tag{2}$$

$$\textbf{curl } \boldsymbol{E} + \frac{\partial \boldsymbol{B}}{\partial t} = \boldsymbol{0}; \quad \textbf{curl } \boldsymbol{E} = -\boldsymbol{E}_0 \times \textbf{grad} \cos \varkappa(z-ct) = (\boldsymbol{E}_0 \times \boldsymbol{k})\varkappa \sin \varkappa(z-ct),$$

$$\frac{\partial \boldsymbol{B}}{\partial t} = c\varkappa \boldsymbol{B}_0 \sin \varkappa(z-ct).$$

Therefore

$$\boldsymbol{E}_0 \times \boldsymbol{k} + c\boldsymbol{B}_0 = \boldsymbol{0}, \quad \text{i.e.} \quad \boldsymbol{k} \times \boldsymbol{E}_0 = c\boldsymbol{B}_0. \tag{3}$$

$$\textbf{curl } \boldsymbol{H} - \frac{\partial \boldsymbol{D}}{\partial t} = \boldsymbol{0}; \quad \textbf{curl } \boldsymbol{H} = -(\boldsymbol{B}_0/\mu_0) \times \textbf{grad} \cos \varkappa(z-ct) = \mu_0^{-1}(\boldsymbol{B}_0 \times \boldsymbol{k})\varkappa \sin \varkappa(z-ct)$$

$$\frac{\partial \boldsymbol{D}}{\partial t} = c\varkappa\varepsilon_0 \boldsymbol{E}_0 \sin \varkappa(z-ct).$$

Therefore

$$\boldsymbol{B}_0 \times \boldsymbol{k} = \mu_0\varepsilon_0 c\boldsymbol{E}_0, \quad \text{i.e.} \quad c(\boldsymbol{B}_0 \times \boldsymbol{k}) = \boldsymbol{E}_0, \tag{4}$$

since $\mu_0\varepsilon_0 = 1/c^2$.

Because the metal is a perfect conductor there can be no fields on the side $z < 0$ of the surface, when the wave is incident from the side $z > 0$. We assume the presence of a reflected ray, but no refracted ray, Fig. 11.4. The charge and current distributions on the metal are such as to "screen" the region in the metal from all fields.

We use the forms used in the first half of the question and write:

Incident ray: $\boldsymbol{E} = \boldsymbol{E}_0 \cos \varkappa(z+ct)$, $\boldsymbol{B} = -\boldsymbol{B}_0 \cos \varkappa(z+ct)$;

Reflected ray: $\boldsymbol{E}' = \boldsymbol{E}_0' \cos \varkappa(z-ct)$, $\boldsymbol{B}' = \boldsymbol{B}_0' \cos \varkappa(z-ct)$.

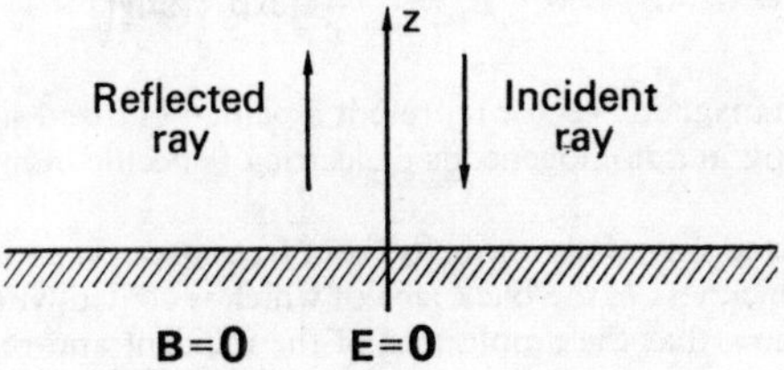

FIG. 11.4

The minus sign occurs in the incident ray since it is travelling in the negative direction of the z-axis. Consequently the relations (3) and (4) become

$$\text{Incident ray:} \quad \boldsymbol{k}\times\boldsymbol{E}_0 = c\boldsymbol{B}_0, \quad c(\boldsymbol{k}\times\boldsymbol{B}_0) = -\boldsymbol{E}_0; \tag{3a}$$

$$\text{Reflected ray:} \quad \boldsymbol{k}\times\boldsymbol{E}_0' = c\boldsymbol{B}_0', \quad c(\boldsymbol{k}\times\boldsymbol{B}_0') = -\boldsymbol{E}_0'. \tag{4a}$$

The boundary conditions take the form, for $z = 0$,

$$\boldsymbol{k}\cdot(\varepsilon_0\boldsymbol{E}+\varepsilon_0\boldsymbol{E}') = \sigma, \quad \boldsymbol{k}\times(\boldsymbol{E}+\boldsymbol{E}') = \boldsymbol{0}, \quad \boldsymbol{k}\cdot(\boldsymbol{B}+\boldsymbol{B}') = 0, \quad \boldsymbol{k}\times(\boldsymbol{B}+\boldsymbol{B}')/\mu_0 = \boldsymbol{K},$$

where σ, $\boldsymbol{K}$ are the surface charge and surface current densities. When we put $z = 0$, these conditions reduce to

$$\boldsymbol{k}\cdot(\boldsymbol{E}_0+\boldsymbol{E}_0')\cos\varkappa ct = \sigma/\varepsilon_0, \quad (5) \qquad \boldsymbol{k}\times(\boldsymbol{E}_0+\boldsymbol{E}_0') = \boldsymbol{0}, \quad (6)$$

$$\boldsymbol{k}\cdot(-\boldsymbol{B}_0+\boldsymbol{B}_0') = 0, \quad (7) \qquad \boldsymbol{k}\times(-\boldsymbol{B}_0+\boldsymbol{B}_0')\cos\varkappa ct = \mu_0\boldsymbol{K} \quad (8)$$

In eqn. (6) we use the relations (3a) and (4a) so that

$$c\boldsymbol{B}_0+c\boldsymbol{B}_0' = \boldsymbol{0}, \quad \text{i.e.} \quad \boldsymbol{B}_0 = -\boldsymbol{B}_0'.$$

We can therefore deduce that

$$\boldsymbol{E}_0' = -c(\boldsymbol{k}\times\boldsymbol{B}_0') = +c(\boldsymbol{k}\times\boldsymbol{B}_0) = -\boldsymbol{E}_0$$

and so $\sigma = 0$ at all times. Because eqn. (2) applies to both reflected and refracted rays (7) is satisfied identically and finally we have

$$\mu_0\boldsymbol{K} = -2(\boldsymbol{k}\times\boldsymbol{B}_0)\cos\varkappa ct,$$

giving the required result.

This current is situated in a field of strength $\boldsymbol{B}+\boldsymbol{B}'$ evaluated for $z = 0$. This field strength is $-2\boldsymbol{B}_0'\cos\varkappa ct$, and so the force exerted on unit area of the conductor is

$$\boldsymbol{K}\times(-2\boldsymbol{B}_0'\cos\varkappa ct) = -(4/\mu_0)([\boldsymbol{B}_0\times\boldsymbol{k}]\times\boldsymbol{B}_0)\cos^2\varkappa ct = -(4/\mu_0)\,\boldsymbol{k}B_0^2\cos^2\varkappa ct.$$

The minus sign shows that the force is into the metal, i.e. a pressure of amount $(4/\mu_0)B_0^2\cos^2\varkappa ct$ having a mean value $2B_0^2/\mu_0$.

Example 2. Prove that the formulae

$$E_x = 0, \quad E_y = A\exp\left\{2\pi i\nu\left(t+\frac{nx}{c}\right)\right\}, \quad E_z = 0$$

for the resolutes of the electric vector and the formulae

$$B_x = 0, \quad B_y = 0, \quad B_z = -\frac{nA}{c}\exp\left\{2\pi i\nu\left(t+\frac{nx}{c}\right)\right\}$$

for the resolutes of the magnetic vector represent a plane-polarized simple harmonic wave of frequency ν travelling in a homogeneous dielectric of specific inductive capacity n^2 and permeability μ_0.

Such a plane-polarized wave falls normally from free space on an infinite slab of homogeneous dielectric of thickness h, the back face of which is coated with a layer of perfectly conducting matter. Show that the amplitudes of the incident and reflected waves in free space are equal, but that, at the front face of the slab, the phase of the reflected wave exceeds that of the incident wave by

$$\pi - 2\tan^{-1}\left(\frac{\tan\theta}{n}\right),$$

where $\theta = 2\pi\nu nh/c$.

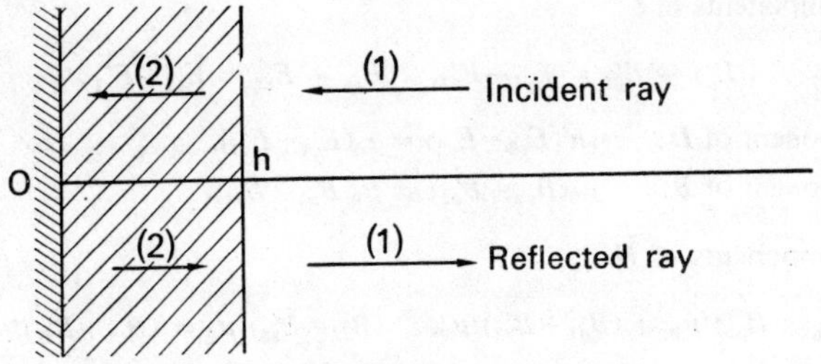

FIG. 11.5

In order to represent the reflections and transmissions that occur we use an incident ray in each section, free space and the dielectric, and corresponding reflected rays, Fig. 11.5. At the conducting plane we apply boundary conditions as in the previous example; further, at the plane $x = h$ we also apply boundary conditions as in (11.24–27).

First we write down suitable expressions for the fields corresponding to the diagram of Fig. 11.5. They are:

Incident ray (1): $E_{x1} = 0,\quad E_{y1} = A\exp\left\{2\pi i\nu\left(t+\frac{x}{c}\right)\right\},\quad E_{z1} = 0,$

$$B_{x1} = 0,\quad B_{y1} = 0,\quad B_{z1} = -\frac{A}{c}\exp\left\{2\pi i\nu\left(t+\frac{x}{c}\right)\right\}.$$

Incident ray (2): $E_{x2} = 0,\quad E_{y2} = A_2\exp\left\{2\pi i\nu\left(t+\frac{nx}{c}\right)\right\},\quad E_{z2} = 0,$

$$B_{x2} = 0,\quad B_{y2} = 0,\quad B_{z2} = -\frac{nA_2}{c}\exp\left\{2\pi i\nu\left(t+\frac{nx}{c}\right)\right\}.$$

Reflected ray (2): $E'_{x2} = 0,\quad E'_{y2} = A'_2\exp\left\{2\pi i\nu\left(t-\frac{nx}{c}\right)\right\},\quad E'_{z2} = 0,$

$$B'_{x2} = 0,\quad B'_{y2} = 0,\quad B'_{z2} = \frac{nA'_2}{c}\exp\left\{2\pi i\nu\left(t-\frac{nx}{c}\right)\right\}.$$

Reflected ray (1): $E'_{x1} = 0,\quad E'_{y1} = A'\exp\left\{2\pi i\nu\left(t-\frac{x}{c}\right)\right\},\quad E'_{z1} = 0,$

$$B'_{x1} = 0,\quad B'_{y1} = 0,\quad B'_{z1} = \frac{A'}{c}\exp\left\{2\pi i\nu\left(t-\frac{x}{c}\right)\right\}.$$

[The expressions differ in the reflected rays from the corresponding incident expressions in the sign of c; otherwise they correspond, with $n = 1$ for region (1).]

At the conductor ($x=0$) the tangential components of **E**, and the normal component of **B**, are zero. The remaining components correspond to the charge and surface currents induced on the conductor by the radiation. These conditions, in terms of components, give

$$E_{y2} + E'_{y2} = 0,\quad E_{z2}+E'_{z2} = 0,\quad B_{x2}+B'_{x2} = 0.$$

These equations lead to the one relation

$$A_2 \exp(2\pi i\nu t) + A_2' \exp(2\pi i\nu t) = 0, \quad \text{i.e.} \quad A_2 + A_2' = 0.$$

At the surface $x = h$ there is no charge or current density so that the boundary conditions give:

Tangential components of $\boldsymbol{E}$:

$$E_{y2} + E_{y2}' = E_{y1} + E_{y1}', \quad E_{z2} + E_{z2}' = E_{z1} + E_{z1}'. \tag{1), (2}$$

Normal component of $\boldsymbol{D}$: $\quad \varepsilon_0 n^2(E_{z2} + E_{z2}') = \varepsilon_0(E_{z1} + E_{z1}').$ (3)

Normal component of $\boldsymbol{B}$: $\quad \mu_0(B_{x2} + B_{x2}') = \mu_0(B_{x1} + B_{x1}').$ (4)

Tangential components of $\boldsymbol{H}$:

$$(B_{y2} + B_{y2}')/\mu_0 = (B_{y1} + B_{y2}')/\mu_0, \quad (B_{z2} + B_{z2}')/\mu_0 = (B_{z1} + B_{z1}')/\mu_0. \tag{5), (6}$$

[In the subsequent work we incorporate the result $A_2 = -A_2'$ already obtained.]

Of these relations only (1) and (6) give non-trivial relations. We obtain the relations by putting $x = h$ so that, from (1),

$$A_2\{\exp(2\pi i\nu nh/c) - \exp(-2\pi i\nu nh/c)\} = A \exp(2\pi i\nu h/c) + A' \exp(-2\pi i\nu h/c),$$

and from (6)

$$-\frac{nA_2}{c}\{\exp(2\pi i\nu nh/c) + \exp(-2\pi i\nu nh/c)\} = -\frac{A}{c}\exp(2\pi i\nu h/c) + \frac{A'}{c}\exp(-2\pi i\nu h/c).$$

These reduce to

$$Ae^{i\theta/n} + A'e^{-i\theta/n} = 2iA_2 \sin\theta, \quad Ae^{i\theta/n} - A'e^{-i\theta/n} = 2nA_2\cos\theta.$$

Therefore

$$\frac{Ae^{i\theta/n}}{A'e^{-i\theta/n}} = -\frac{n\cos\theta + i\sin\theta}{n\cos\theta - i\sin\theta}. \tag{7}$$

Hence $|A| = |A'|$, showing that the amplitudes of the incident and reflected rays in free space are equal. At the front face of the slab in free space the incident and reflected rays are given by the (complex) expressions

$$Ae^{i\theta/n}\exp(2\pi i\nu t), \quad Ae^{-i\theta/n}\exp(2\pi i\nu t).$$

Hence the phase difference between these two waves is the difference between those of $Ae^{i\theta/n}$, $A'e^{-i\theta/n}$. From eqn. (7),

$$A'e^{-i\theta/n} = -Ae^{i\theta/n}e^{-2i\delta},$$

where $\tan\delta = (\tan\theta)/n$. Thus the phase of the reflected ray exceeds that of the incident ray by

$$\pi - 2\delta = \pi - 2\tan^{-1}\left(\frac{\tan\theta}{n}\right).$$

Example 3. Show that $\boldsymbol{E}$, $\boldsymbol{B}$ defined by

$$E_x = Ae^{-i\omega(t-uz)}, \quad E_y = E_z = 0, \quad B_x = 0, \quad cB_y = \sqrt{(K)}Ae^{-i\omega(t-uz)}, \quad B_z = 0,$$

where A, ω are constants, satisfy Maxwell's equations for a uniform medium of dielectric constant K and permeability μ_0, provided that $u = \sqrt{K}/c$.

The region $0 \leqslant z \leqslant h$ is filled with material of dielectric constant 4 and permeability μ_0, and the rest of space is empty. A plane-polarized plane wave of frequency $\omega/2\pi$, travelling in the positive direction of Oz, is normally incident on the face $z = 0$. Show that the ratio of the amplitude of this incident wave to the amplitude of the reflected wave in $z < 0$ is

$$(1+\tfrac{16}{9}\operatorname{cosec}^2\theta)^{1/2},$$

where $\theta = 2\omega h/c$.

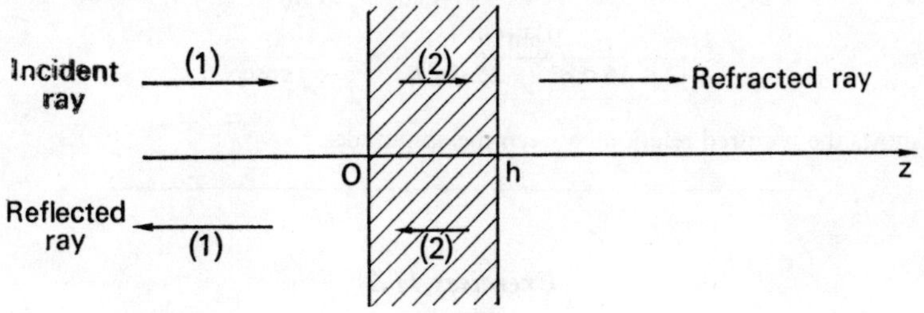

FIG. 11.6

The expressions quoted can easily be shown to satisfy Maxwell's equations. Since reflection and transmission can occur at both surfaces we assume a set of incident rays as shown in the diagram of Fig. 11.6. Adopting the notation of the first part of the question, we use the following expressions.

Incident ray (1): $E_{x1} = Ae^{-i\omega(t-z/c)},\ E_{y1} = 0,\ E_{z1} = 0,$
$B_{x1} = 0,\ B_{y1} = (A/c)e^{-i\omega(t-z/c)},\ B_{z1} = 0.$ $\quad z \leqslant 0$

Incident ray (2): $E_{x2} = A_2e^{-i\omega(t-2z/c)},\ E_{y2} = 0,\ E_{z2} = 0,$
$B_{x2} = 0,\ B_{y2} = (2A_2/c)e^{-i\omega(t-2z/c)},\ B_{z2} = 0.$ $\quad 0 \leqslant z \leqslant h$

Refracted ray: $E''_x = A''e^{-i\omega(t-z/c)},\ E''_y = 0,\ E''_z = 0,$
$B''_x = 0,\ B''_y = (A''/c)e^{-i\omega(t-z/c)},\ B''_z = 0.$ $\quad z \geqslant 0$

Reflected ray (2): $E'_{x2} = A'_2e^{-i\omega(t+2z/c)},\ E'_{y2} = 0,\ E'_{z2} = 0,$
$B'_{x2} = 0,\ B'_{y2} = -(2A'_2/c)e^{-i\omega(t+2z/c)},\ B'_{z2} = 0.$ $\quad 0 \leqslant z \leqslant h$

Reflected ray (1): $E'_{x1} = A'e^{-i\omega(t+z/c)},\ E'_{y1} = 0,\ E'_{z1} = 0,$
$B'_{x1} = 0,\ B'_{y1} = -(A'/c)e^{-i\omega(t+z/c)},\ B'_{z1} = 0.$ $\quad z \leqslant 0$

The boundary conditions to be applied at each surface are the standard ones corresponding to no charge or current on either surface. As in previous cases, only a few of the component equations give non-trivial relations.

From the conditions at $z = 0$ we obtain:
tangential components of $\boldsymbol{E}$:

$$Ae^{-i\omega t}+A'e^{-i\omega t} = A_2e^{-i\omega t}+A'_2e^{-i\omega t},$$

tangential component of $\boldsymbol{B}$:

$$(A/c)e^{-i\omega t}-(A'/c)e^{-i\omega t} = (2A_2/c)e^{-i\omega t}-(2A'_2/c)e^{-i\omega t},$$

i.e. $\quad A+A' = A_2+A'_2, \quad A-A' = 2(A_2-A'_2),$

whence $\quad A = (3A_2-A'_2)/2, \quad A' = (-A_2+3A'_2)/2. \quad (1)$

From the conditions at $z = h$ we obtain:

tangential components of $\boldsymbol{E}$: $\quad A_2 e^{2i\omega h/c} + A_2' e^{-2i\omega h/c} = A'' e^{i\omega h/c}$, (2)

tangential component of $\boldsymbol{B}$: $\quad 2A_2 e^{2i\omega h/c} - 2A_2' e^{-2i\omega h/c} = A'' e^{i\omega h/c}$. (3)

We eliminate the quantities A_2, A_2', A'' by solving (2) and (3) for A_2, A_2' and then substituting into eqn. (1). After some manipulation we obtain

$$\frac{A'}{A} = \frac{3e^{i\theta} - 3e^{-i\theta}}{-e^{i\theta} + 9e^{-i\theta}} = \frac{3i \sin\theta}{4\cos\theta - 5i\sin\theta}.$$

Hence
$$\left|\frac{A'}{A}\right|^2 = \frac{9\sin^2\theta}{16\cos^2\theta + 25\sin^2\theta} = \frac{1}{1 + \frac{16}{9}\operatorname{cosec}^2\theta}.$$

This shows the required relation between the amplitudes.

Exercises 11.2

1. Show that a solution of the equations of the electromagnetic field in a vacuum is given by $\boldsymbol{B} = \nabla \times \boldsymbol{A}$ and $\boldsymbol{E} = -\dot{\boldsymbol{A}}$, where $\nabla \cdot \boldsymbol{A} = 0$, $\nabla^2 \boldsymbol{A} - (1/c^2)\ddot{\boldsymbol{A}} = 0$. Verify that these conditions are satisfied by taking $\mathbf{A}$ to be

$$\{\cos(\omega t + kz) \sin(\omega t + kz)\ 0\},$$

where $kc = \omega$; determine the corresponding vectors $\boldsymbol{B}$ and $\boldsymbol{E}$.

A wave of this sort in the region $z > 0$ falls upon the face $z = 0$ of a perfect conductor. Find the reflected wave, and show that, at any fixed point, the vectors $\boldsymbol{B}$ and $\boldsymbol{E}$ for the combined field of the incident and reflected waves are parallel and constant in magnitude.

2. Space is empty except for a uniform slab of dielectric of unit permeability and dielectric constant n^2, and of thickness h. A polarized plane electromagnetic wave of vacuum wavelength $2nh$ is incident normally on the slab. Show that outside the slab there is no reflected wave and that in the centre of the slab the amplitude of the electric field is $1/n$ of its amplitude outside the slab.

3. A plane polarized wave of period $2\pi/n$ travels in free space and is normally incident on a plane slab of thickness h and dielectric constant q^2. Prove that the intensity $\boldsymbol{I}$ of the transmitted wave is related to the intensity I_0 of the incident wave by

$$\frac{I_0}{I} = 1 + \frac{1}{4}\left(q - \frac{1}{q}\right)^2 \sin^2\theta,$$

where $\theta = qnh/c$.

4. Show that a solution of Maxwell's equations for the electromagnetic field in a non-magnetic insulator with dielectric constant K is given by

$$E_x = A \exp i(\omega t - \alpha z), \qquad H_y = A(c\alpha/\omega) \exp i(\omega t - \alpha z),$$

where A is constant and the remaining components are zero, provided that $c\alpha = \omega\sqrt{K}$.

Three such media, with dielectric constants K_1, K_2, K_3, occupy the regions $z < 0$, $0 < z < d$, $d < z$ respectively. There is an incident wave of the above form in the first medium. If the thickness of the second medium in one-quarter of a wavelength, so that

$\alpha_2 d = \frac{1}{2}\pi$ (where $c\alpha_2 = \omega\sqrt{K_2}$), show that the ratio of the amplitudes of the reflected wave and the incident wave in the region $z < 0$ is

$$\left|\frac{\sqrt{(K_1K_3)}-K_2}{\sqrt{(K_1K_3)}+K_2}\right|.$$

11.3 Reflection and refraction: oblique incidence

To discuss the process of reflection and refraction at oblique incidence in terms of Maxwell's equations we now need to use plane harmonic wave solutions of Maxwell's equations whose direction of propagation is inclined to the normal. We use a coordinate frame of axes which has the positive direction of the x-axis directed into the optically less dense medium, i.e. the same frame as we used in the last section; and so we must use waves which are propagated in a direction specified by the unit vector $\hat{\boldsymbol{n}}$, which is not parallel to a coordinate axis.

For a wave propagated in this direction the space variable x in (11.12) and (11.13) must be replaced by $\hat{\boldsymbol{n}}\cdot\boldsymbol{r}$, where $\boldsymbol{r}$ is the position vector of a field point. When the wave is a plane polarized wave of frequency $\omega/(2\pi)$ we can take the expression for $\boldsymbol{B}$ as

$$\boldsymbol{B} = \boldsymbol{b}\exp \mathrm{i}\{\omega t - k(\hat{\boldsymbol{n}}\cdot\boldsymbol{r})\}, \tag{11.36}$$

where $\boldsymbol{b}$ is a constant vector. Since $\boldsymbol{E}$, $\boldsymbol{B}$, $\hat{\boldsymbol{n}}$ form a right-handed triad we then have

$$\boldsymbol{E} = c(\boldsymbol{b}\times\hat{\boldsymbol{n}})\exp \mathrm{i}\{\omega t - k(\hat{\boldsymbol{n}}\cdot\boldsymbol{r})\} \tag{11.37}$$

for the electric field. The velocity of propagation c is given by [eqn. (11.16)]

$$c = \omega/k.$$

Taking directions of reflected and refracted rays as indicated in Fig. 11.2 we assume the following forms:

$$\text{reflected ray:}\quad \begin{aligned} \boldsymbol{B}' &= \boldsymbol{b}'\exp \mathrm{i}\{\omega' t - k'(\hat{\boldsymbol{n}}'\cdot\boldsymbol{r})\},\\ \boldsymbol{E}' &= c(\boldsymbol{b}'\times\hat{\boldsymbol{n}}')\exp \mathrm{i}\{\omega' t - k'(\hat{\boldsymbol{n}}'\cdot\boldsymbol{r})\}; \end{aligned} \tag{11.38}$$

$$\text{refracted ray:}\quad \begin{aligned} \boldsymbol{B}'' &= \boldsymbol{b}''\exp \mathrm{i}\{\omega'' t - k''(\hat{\boldsymbol{n}}''\cdot\boldsymbol{r})\},\\ \boldsymbol{E}'' &= v(\boldsymbol{b}''\times\hat{\boldsymbol{n}}'')\exp \mathrm{i}\{\omega'' t - k''(\hat{\boldsymbol{n}}''\cdot\boldsymbol{r})\}. \end{aligned} \tag{11.39}$$

In these expressions the velocities of propagation are given by

$$c = \omega'/k', \quad v = \omega''/k'' = c/\sqrt{K}. \tag{11.22}$$

All these expressions can be shown to satisfy Maxwell's equations in the respective regions.

The conditions which must be satisfied at the boundary $x = 0$ are the same as those set out in the previous section in eqns. (11.23) to (11.27), viz.

$$\boldsymbol{B}_+ = \boldsymbol{B}+\boldsymbol{B}', \quad \boldsymbol{E}_+ = \boldsymbol{E}+\boldsymbol{E}' \quad \text{for} \quad x > 0,$$
$$\boldsymbol{B}_- = \boldsymbol{B}'', \quad \boldsymbol{E}_- = \boldsymbol{E}'' \quad \text{for} \quad x < 0;$$

and, at the surface of discontinuity,

$$\boldsymbol{i}\cdot(\boldsymbol{B}_+ - \boldsymbol{B}_-) = 0, \quad \boldsymbol{i}\times(\boldsymbol{B}_+ - \boldsymbol{B}_-) = \boldsymbol{0},$$
$$\boldsymbol{i}\cdot(\varepsilon_0 \boldsymbol{E}_+ - \varepsilon_0 K \boldsymbol{E}_-) = 0, \quad \boldsymbol{i}\times(\boldsymbol{E}_+ - \boldsymbol{E}_-) = \boldsymbol{0}.$$

For the same reason as before, in order to ensure that these conditions are satisfied at all times we must have

$$\omega = \omega' = \omega''.$$

Now, however, when we put $x = 0$ on the boundary the exponential factors nvolve y and z in the following way:

incident ray: $\exp \mathrm{i}\{-k\hat{\boldsymbol{n}}\cdot(\boldsymbol{j}y+\boldsymbol{k}z)\}$;
reflected ray: $\exp \mathrm{i}\{-k'\hat{\boldsymbol{n}}'\cdot(\boldsymbol{j}y+\boldsymbol{k}z)\}$;
refracted ray: $\exp \mathrm{i}\{-k''\hat{\boldsymbol{n}}''\cdot(\boldsymbol{j}y+\boldsymbol{k}z)\}$.

The boundary conditions have to be satisfied at every point on the surface $x = 0$ and so each of these exponential factors must be the same, and we obtain the following results by equating coefficients of y and z,

$$k(\hat{\boldsymbol{n}}\cdot\boldsymbol{j}) = k'(\hat{\boldsymbol{n}}'\cdot\boldsymbol{j}) = k''(\hat{\boldsymbol{n}}''\cdot\boldsymbol{j}), \tag{11.40}$$
$$k(\hat{\boldsymbol{n}}\cdot\boldsymbol{k}) = k'(\hat{\boldsymbol{n}}'\cdot\boldsymbol{k}) = k''(\hat{\boldsymbol{n}}''\cdot\boldsymbol{k}). \tag{11.41}$$

Since the orientation of the y- and z-axes is, as yet, unspecified we now choose the z-axis to be perpendicular to the plane of incidence, i.e. $\hat{\boldsymbol{n}}\times\boldsymbol{i}$ is parallel to $\boldsymbol{k}$, or $\hat{\boldsymbol{n}}\cdot\boldsymbol{k} = 0$. Equation (11.41) now shows that the reflected and refracted rays also lie in this plane. We take the plane of the diagram of Fig. 11.7 to be the "plane of incidence" in which the dotted lines represent the wave-fronts of the respective rays. When we substitute for k, k', k'' in eqn. (11.40) from eqns. (11.16) and (11.22) we obtain

$$\omega(\hat{\boldsymbol{n}}\cdot\boldsymbol{j})/c = \omega(\hat{\boldsymbol{n}}'\cdot\boldsymbol{j})/c = \omega\sqrt{K}(\hat{\boldsymbol{n}}''\cdot\boldsymbol{j})/c.$$

Therefore

$$\hat{\boldsymbol{n}}\cdot\boldsymbol{j} = \hat{\boldsymbol{n}}'\cdot\boldsymbol{j}, \quad \text{i.e.} \quad \sin\theta = \sin\theta', \tag{11.42}$$
$$\hat{\boldsymbol{n}}''\cdot\boldsymbol{j} = (\hat{\boldsymbol{n}}\cdot\boldsymbol{j})/\sqrt{K}, \quad \text{i.e.} \quad \sin\theta'' = \sin\theta/\sqrt{K}. \tag{11.43}$$

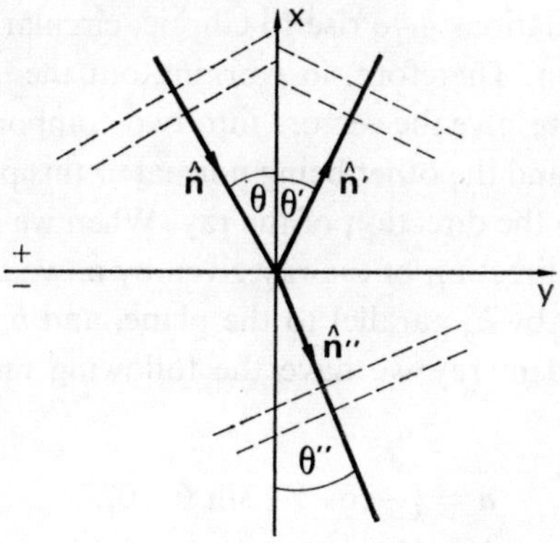

FIG. 11.7

Hence the refractive index is given by

$$n^2 = K \tag{11.44}$$

and the laws of reflection and refraction are verified. Since $v = c/\sqrt{K}$ we see that the refractive index is also given by

$$n = c/v. \tag{11.45}$$

Thus far the vectors $\boldsymbol{b}$, $\boldsymbol{b}'$, $\boldsymbol{b}''$ have not entered into the calculations; the remaining steps in working out the implications of the boundary conditions at the line $x = 0$ give the relations between the amplitudes of the incident, reflected and refracted rays, corresponding to the results in eqn. (11.34) for normal incidence. When we substitute the expressions (11.36) to (11.39) into the boundary conditions, the exponential factors cancel out and we are left with four relations which must be satisfied by the amplitudes $\boldsymbol{b}$, $\boldsymbol{b}'$, $\boldsymbol{b}''$. These relations are:

$$\boldsymbol{i}\cdot(\boldsymbol{B}_+ - \boldsymbol{B}_-) = 0; \qquad \boldsymbol{i}\cdot(\boldsymbol{b}+\boldsymbol{b}'-\boldsymbol{b}'') = 0; \tag{11.46}$$

$$\boldsymbol{i}\times(\boldsymbol{B}_+ - \boldsymbol{B}_-) = \boldsymbol{0}; \qquad \boldsymbol{i}\times(\boldsymbol{b}+\boldsymbol{b}'-\boldsymbol{b}'') = \boldsymbol{0}; \tag{11.47}$$

$$\boldsymbol{i}\cdot(\varepsilon_0\boldsymbol{E}_+ - K\varepsilon_0\boldsymbol{E}_-) = 0; \qquad \boldsymbol{i}\cdot\{c(\boldsymbol{b}\times\hat{\boldsymbol{n}}) + c(\boldsymbol{b}'\times\hat{\boldsymbol{n}}') - vK(\boldsymbol{b}''\times\hat{\boldsymbol{n}}'')\} = 0; \tag{11.48}$$

$$\boldsymbol{i}\times(\boldsymbol{E}_+ - \boldsymbol{E}_-) = \boldsymbol{0}; \qquad \boldsymbol{i}\times\{c(\boldsymbol{b}\times\hat{\boldsymbol{n}}) + c(\boldsymbol{b}'\times\hat{\boldsymbol{n}}') - v(\boldsymbol{b}''\times\hat{\boldsymbol{n}}'')\} = \boldsymbol{0}. \tag{11.49}$$

In order to make progress in working out the implications of these equations we must take account of the direction of polarization of the fields. The reflected and refracted waves depend upon the direction of polarization of the incident wave. In § 11.1 we saw that the magnetic vector consisted of two components at right angles, each component corresponding to a plane-polarized wave. Combinations of two such components with various amplitudes and phase relations give rise to elliptic, circular, left-handed or right-handed polarization. Therefore, in working out the consequences of eqns. (11.46)–(11.49) we resolve the vectors into two components, one lying in the plane of incidence and the other being normal to this plane; each component is at right angles to the direction of the ray. When we resolve the vector $\boldsymbol{b}$ in this way, with the direction of the ray given by $\hat{\boldsymbol{n}}$, we denote the magnitudes of the components by b_p parallel to the plane, and b_n normal to the plane. Thus, for the incident ray we have the following resolutions, in the *xyz*-frame:

Incident ray:
$$\hat{\boldsymbol{n}} = \{-\cos\theta \quad \sin\theta \quad 0\}, \tag{11.50}$$
$$\boldsymbol{b} = \{b_p \sin\theta \quad b_p \cos\theta \quad b_n\}, \tag{11.51}$$
$$\boldsymbol{b}\times\hat{\boldsymbol{n}} = \{-b_n \sin\theta \quad -b_n \cos\theta \quad b_p\}. \tag{11.52}$$

The same resolution applied to the other rays gives

Reflected ray:
$$\hat{\boldsymbol{n}}' = \{\cos\theta \quad \sin\theta \quad 0\}, \tag{11.53}$$
$$\boldsymbol{b}' = \{b_p' \sin\theta \quad -b_p' \cos\theta \quad b_n'\}, \tag{11.54}$$
$$\boldsymbol{b}'\times\hat{\boldsymbol{n}}' = \{-b_n' \sin\theta \quad b_n' \cos\theta \quad b_p'\}. \tag{11.55}$$

Refracted ray:
$$\hat{\boldsymbol{n}}'' = \{-\cos\theta'' \quad \sin\theta'' \quad 0\}, \tag{11.56}$$
$$\boldsymbol{b}'' = \{b_p'' \sin\theta'' \quad b_p'' \cos\theta'' \quad b_n''\}, \tag{11.57}$$
$$\boldsymbol{b}''\times\hat{\boldsymbol{n}}'' = \{-b_n'' \sin\theta'' \quad -b_n'' \cos\theta'' \quad b_p''\}. \tag{11.58}$$

We can now use eqns. (11.46)–(11.49) to determine the amplitudes b_p', b_n', b_p'', b_n'' in terms of those for the incident ray, viz. b_p, b_n. We obtain the following relations on taking the components of eqns. (11.46)–(11.49):

(11.46) gives:
$$b_p \sin\theta + b_p' \sin\theta - b_p'' \sin\theta'' = 0; \tag{11.59}$$

(11.47) gives:
$$b_p \cos\theta - b_p' \cos\theta + b_p'' \cos\theta'' = 0, \tag{11.60}$$
$$b_n + b_n' - b_n'' = 0; \tag{11.61}$$

(11.48) gives:
$$-cb_n' \sin\theta - cb_n' \sin\theta + vKb_n'' \sin\theta'' = 0; \tag{11.62}$$

(11.49) gives:
$$-cb_n \cos\theta + cb_n' \cos\theta - vb_n'' \cos\theta'' = 0, \tag{11.63}$$
$$cb_p + cb_p' - vb_p'' = 0. \tag{11.64}$$

Because of Snell's law and the relation between the velocities, the refractive index and K given in eqns. (11.43), (11.44), (11.45), we see that eqns. (11.61) and (11.62) are identical, and eqns. (11.59) and (11.64) are identical also. Thus eqns. (11.59)–(11.64) reduce to two pairs of equations; one pair relates b_p', b_p'' to b_p, and the other pair relates b_n', b_n'' to b_n. In fact the components polarized parallel and normal to the plane of incidence behave independently of each other. The relations (Fresnel's formulae) are:

for b_p', b_p'':

$$cb_p' \quad -vb_p'' \quad +cb_p \quad = 0,$$
$$-b_p' \cos\theta + b_p'' \cos\theta'' + b_p \cos\theta = 0;$$

$$\therefore \qquad b_p' = \frac{c\cos\theta'' - v\cos\theta}{c\cos\theta'' + v\cos\theta} b_p, \quad b_p'' = \frac{2\cos\theta}{c\cos\theta'' + v\cos\theta} b_p; \tag{11.65}$$

and for b_n', b_n''

$$b_n' \quad -b_n'' \quad +b_n \quad = 0,$$
$$cb_n' \cos\theta - vb_n'' \cos\theta'' - cb_n \cos\theta = 0;$$

$$\therefore \qquad b_n' = \frac{c\cos\theta - v\cos\theta''}{c\cos\theta + v\cos\theta''} b_n, \quad b_n'' = \frac{2c\cos\theta}{c\cos\theta + v\cos\theta''} b_n. \tag{11.66}$$

Note that when $\theta = 0$ eqns. (11.65) and (11.66) both reduce to the result (11.33) obtained for normal incidence. These equations can be expressed alternatively, after use of Snell's law, in the form

$$\frac{c}{\sin\theta} = \frac{v}{\sin\theta''},$$

thus

$$b_p' = \frac{\sin(\theta - \theta'')}{\sin(\theta + \theta'')} b_p, \quad b_p'' = \frac{\sin 2\theta}{\sin(\theta + \theta'')} b_p, \tag{11.65a}$$

and

$$b_n' = \frac{\tan(\theta - \theta'')}{\tan(\theta + \theta'')} b_n, \quad b_n'' = \frac{\sin 2\theta}{\sin(\theta + \theta'')\cos(\theta - \theta'')} b_n. \tag{11.66a}$$

The resultant fields can now be written out in the two cases.

(a) *Magnetic vector normal to the plane of incidence*, i.e. $b_p = 0$ so that $\boldsymbol{l} = \boldsymbol{k}b_n$. Then for the incident ray:

$$\begin{aligned} \boldsymbol{B}_a &= \boldsymbol{k}b_n \exp i\{\omega t - k(\hat{\boldsymbol{n}}\cdot\boldsymbol{r})\}, \\ \boldsymbol{E}_a &= c(\boldsymbol{k}\times\hat{\boldsymbol{n}})b_n \exp i\{\omega t - k(\hat{\boldsymbol{n}}\cdot\boldsymbol{r})\} = c(\boldsymbol{B}_a\times\hat{\boldsymbol{n}}); \end{aligned} \tag{11.67}$$

for the reflected ray:

$$\begin{aligned} \boldsymbol{B}_a' &= \boldsymbol{k}b_n' \exp i\{\omega t - k'(\hat{\boldsymbol{n}}'\cdot\boldsymbol{r})\}, \\ \boldsymbol{E}_a' &= c(\boldsymbol{k}\times\hat{\boldsymbol{n}}')b_n' \exp i\{\omega t - k'(\hat{\boldsymbol{n}}'\cdot\boldsymbol{r})\} = c(\boldsymbol{B}_a'\times\hat{\boldsymbol{n}}'); \end{aligned} \tag{11.68}$$

for the refracted ray:

$$\begin{aligned} \boldsymbol{B}_a'' &= \boldsymbol{k}b_n'' \exp i\{\omega t - k''(\hat{\boldsymbol{n}}''\cdot\boldsymbol{r})\}, \\ \boldsymbol{E}_a'' &= c(\boldsymbol{k}\times\hat{\boldsymbol{n}}'')b_n'' \exp i\{\omega t - k''(\hat{\boldsymbol{n}}''\cdot\boldsymbol{r})\} = v(\boldsymbol{B}_a''\times\hat{\boldsymbol{n}}''). \end{aligned} \tag{11.69}$$

(b) *Magnetic vector in the plane of incidence*, i.e. $b_n = 0$. In this case the electric vector is normal to the plane of incidence so that we have the following results:

Incident ray:

$$\begin{aligned} \boldsymbol{B}_b &= (\hat{\boldsymbol{n}}\times\boldsymbol{k})b_p \exp \mathrm{i}\{\omega t - k(\hat{\boldsymbol{n}}\cdot\boldsymbol{r})\} = (\hat{\boldsymbol{n}}\times\boldsymbol{E}_b)/c, \\ \boldsymbol{E}_b &= ckb_p \exp \mathrm{i}\{\omega t - k(\hat{\boldsymbol{n}}\cdot\boldsymbol{r})\}, \end{aligned} \tag{11.70}$$

Reflected ray:

$$\begin{aligned} \boldsymbol{B}'_b &= (\hat{\boldsymbol{n}}'\times\boldsymbol{k})b'_p \exp \mathrm{i}\{\omega t - k'(\hat{\boldsymbol{n}}'\cdot\boldsymbol{r})\} = (\hat{\boldsymbol{n}}\times\boldsymbol{E}'_b)/c, \\ \boldsymbol{E}'_b &= ckb'_p \exp \mathrm{i}\{\omega t - k'(\hat{\boldsymbol{n}}'\cdot\boldsymbol{r})\}, \end{aligned} \tag{11.71}$$

Refracted ray:

$$\begin{aligned} \boldsymbol{B}''_b &= (\hat{\boldsymbol{n}}''\times\boldsymbol{k})b''_p \exp \mathrm{i}\{\omega t - k''(\hat{\boldsymbol{n}}''\cdot\boldsymbol{r})\} = (\hat{\boldsymbol{n}}''\times\boldsymbol{E}''_b)/v, \\ \boldsymbol{E}''_b &= vkb''_p \exp \mathrm{i}\{\omega t - k''(\hat{\boldsymbol{n}}''\cdot\boldsymbol{r})\}. \end{aligned} \tag{11.72}$$

These formulae also prove another well-known physical result, Brewster's law. If, in eqn. (11.66a), the angles θ, θ'' satisfy

$$\theta + \theta'' = \pi/2, \tag{11.73}$$

then $b'_n = 0$, $b'_p \neq 0$. Hence, when eqn. (11.73) is satisfied so that the reflected and refracted rays are perpendicular, the reflected ray is polarized in one plane. The angle of incidence, θ, for which this occurs is called the polarizing angle and is given by

$$n = \frac{\sin\theta}{\sin\theta''} = \frac{\sin\theta}{\cos\theta} = \tan\theta. \tag{11.74}$$

For other angles of incidence, provided that b_p and b_n are of comparable orders of magnitude, this result means that $b'_p > b'_n$ and the reflected light is predominantly composed of light polarized in one plane. If this light is viewed through spectacles which do not allow light polarized in this plane to pass, then the glare caused by light reflected from a surface is considerably reduced. This is the principle behind the action of polaroid spectacles which reduce the glare of reflected light off a shiny road surface or the sea.

11.4 Energy relations for oblique incidence

In discussions of energy in connection with harmonic waves the results are simplified by using mean values which are easily obtained in the complex number representation we are using. We may assume that *one* of the amplitude components of the incident ray is real; then for elliptic or circular polarization the other component differs in its argument as well as its amplitude from the first (see § 11.1, p. 444).

The mean rate of propagation of energy in connection with a harmonic wave such as that given by eqns. (11.36) and (11.37) is the real part of $\boldsymbol{P} = \frac{1}{2}(\boldsymbol{E}\times\boldsymbol{H}^*) = (\boldsymbol{E}\times\boldsymbol{B}^*)/(2\mu_0)$. Therefore,

$$\boldsymbol{P} = c(\boldsymbol{b}\times\hat{\boldsymbol{n}})\times\boldsymbol{b}^*/(2\mu_0).$$

If we consider an incident ray given by eqns. (11.50–52),

$$\boldsymbol{P} = \left(\frac{c}{2\mu_0}\right)\{\hat{\boldsymbol{n}}(\boldsymbol{b}\cdot\boldsymbol{b}^*)-\boldsymbol{b}(\hat{\boldsymbol{n}}\cdot\boldsymbol{b}^*)\}.$$

But $$\boldsymbol{b}\cdot\boldsymbol{b}^* = b_p b_p^* + b_n b_n^*, \quad \hat{\boldsymbol{n}}\cdot\boldsymbol{b} = 0 = \hat{\boldsymbol{n}}\cdot\boldsymbol{b}^*.$$

Therefore $$\boldsymbol{P} = \frac{c}{2\mu_0}\{|b_p|^2+|b_n|^2\}\,\hat{\boldsymbol{n}}. \tag{11.75}$$

Similarly there are energy flows associated with the reflected and refracted rays given by

$$\boldsymbol{P}' = \frac{c}{2\mu_0}(|b_p'|^2+|b_n'|^2)\,\hat{\boldsymbol{n}}', \tag{11.76}$$

and

$$\boldsymbol{P}'' = \frac{v}{2\mu_0}(|b_p''|^2+|b_n''|^2)\,\hat{\boldsymbol{n}}''. \tag{11.77}$$

To investigate the pattern of energy flow on each side of the surface of separation we consider the two vectors

$$\boldsymbol{P}_+ = \tfrac{1}{2}\boldsymbol{E}_+\times\boldsymbol{H}_+^* = (\boldsymbol{E}+\boldsymbol{E}')\times(\boldsymbol{B}^*+\boldsymbol{B}'^*)/(2\mu_0)$$

and

$$\boldsymbol{P}_- = \tfrac{1}{2}\boldsymbol{E}_-\times\boldsymbol{H}_-^* = \boldsymbol{P}''.$$

The flow in the positive region ($x > 0$) is given by the real part of

$$\boldsymbol{P}_+ = \boldsymbol{P}+\boldsymbol{P}'+(\boldsymbol{E}\times\boldsymbol{B}'^*+\boldsymbol{E}'\times\boldsymbol{B}^*)/(2\mu_0). \tag{11.78}$$

Although it is not obvious at first sight, substitution of the various expressions in (11.36–8) for the field vectors shows that the last term in (11.78) is an imaginary quantity, so that the energy flow above the surface of separation reduces to the sum of the contributions from each ray separately. The flow of energy in the normal direction toward the surface in the positive region is

$$-\boldsymbol{P}_+\cdot\boldsymbol{i} = c\cos\theta(|b_p|^2+|b_n|^2-|b_p'|^2-|b_n'|^2)/(2\mu_0) \tag{11.79}$$

and the flow in the negative region away from the surface is

$$-\boldsymbol{P}_{-}\cdot\boldsymbol{i} = v\cos\theta''(|b_p''|^2+|b_n''|^2)/(2\mu_0). \tag{11.80}$$

The relations obtained from the boundary conditions [used best in the forms (11.65) and (11.66)] show that the energies involving b_p, b_p', b_p'', and the energies involving b_n, b_n', b_n'' balance separately in (11.79) and (11.80). Hence the two modes of polarization discussed on pp. 459–460 of the previous section behave independently, and the energy brought to the surface by the incident ray, for either polarization, is carried away from the surface by the reflected and refracted rays of the same polarization. The components of $\boldsymbol{P}_+$, $\boldsymbol{P}_-$ in directions parallel to the surface are not equal; this means that the oblique rays give a net flow of energy both parallel to the surface of separation and normal to it. It is only the latter components which, through eqns. (11.79) and (11.80), need to satisfy conservation of energy.

We define coefficients of reflection and transmission for these oblique rays, as in § 11.2, but here we have one of each coefficient for each mode of polarization, four coefficients in all. They are:

(a) *Magnetic vector normal to the plane of incidence:*

$$R_n = \frac{\text{Reflected energy flow}}{\text{Incident energy flow}} = \frac{|b_n'|^2\, c\cos\theta}{|b_n|^2\, c\cos\theta} = \frac{|b_n'|^2}{|b_n|^2}$$

$$= \frac{\tan^2(\theta-\theta'')}{\tan^2(\theta+\theta'')}, \tag{11.81}$$

$$T_n = \frac{\text{Refracted energy flow}}{\text{Incident energy flow}} = \frac{|b_n''|^2\, v\cos\theta''}{|b_n|^2\, c\cos\theta}$$

$$= \frac{2\sin 2\theta\cos\theta''\sin\theta''}{\sin^2(\theta+\theta'')\cos^2(\theta-\theta'')}. \tag{11.82}$$

(b) *Magnetic vector parallel to the plane of incidence:*

$$R_p = \frac{\text{Reflected energy flow}}{\text{Incident energy flow}} = \frac{|b_p'|^2\, c\cos\theta}{|b_p|^2\, c\cos\theta} = \frac{|b_p'|^2}{|b_p|^2}$$

$$= \frac{\sin^2(\theta-\theta'')}{\sin^2(\theta+\theta'')}, \tag{11.83}$$

$$T_p = \frac{\text{Refracted energy flow}}{\text{Incident energy flow}} = \frac{|b_p''|^2\, v\cos\theta''}{|b_p|^2\, c\cos\theta}$$

$$= \frac{2\sin 2\theta\sin\theta''\cos\theta''}{\sin^2(\theta+\theta'')}. \tag{11.84}$$

In these expressions we have eliminated the velocities v and c by using Snell's law of refraction in the form $v \sin\theta = c \sin\theta''$. It can easily be verified that $R_n+T_n = 1$, and $R_p+T_p = 1$.

These coefficients are different, for a given angle of incidence, for the two modes of polarization and so it follows that the polarizations of the reflected and refracted rays are both altered from that of the incident ray by the encounter with the surface, and these differences, as well as the intensities of the rays, vary with the angle of incidence.

Exercises 11.4

1. State the conditions under which a linearly polarized plane electromagnetic wave will be totally reflected at a plane interface separating two dielectric media of dielectric constants K_1 and K_2 and permeability μ_0. Obtain an expression, in terms of K_1, K_2 and the angle of incidence θ, for the phase difference δ between the components in and normal to the plane of incidence of the electric vector of the reflected wave, and show that δ cannot exceed the value

$$2\tan^{-1}\left(\frac{1-n^2}{2n}\right),$$

where $n = \sqrt{(K_2/K_1)}$.

2. A plane electromagnetic wave, polarized with the electric vector in the plane of incidence, crosses a plane boundary from a medium with dielectric constant K_1 to a medium with dielectric constant K_2, the magnetic permeability being μ_0 for each. If θ, ϕ are the angles of incidence and refraction, respectively, prove that the angle of reflection is θ, and that $K_1^{1/2}\sin\theta = K_2^{1/2}\sin\phi$.

Prove also that the amplitudes E_1, E', E_2, of the incident, reflected, and refracted fields are related by

$$\frac{E'}{E_1} = \frac{\tan(\theta-\phi)}{\tan(\theta+\phi)}, \qquad \frac{E_2}{E_1} = \frac{2\sin\phi\cos\theta}{\sin(\theta+\phi)\cos(\theta-\phi)}.$$

3. The field vectors of a train of plane polarized electromagnetic waves travelling in a medium of dielectric constant k and permeability μ_0 at a field point of position vector $\boldsymbol{r}$ are given by

$$\boldsymbol{E} = \boldsymbol{A}\exp\{i\omega(t-\boldsymbol{u}\cdot\boldsymbol{r})\},$$
$$\boldsymbol{H} = \boldsymbol{C}\exp\{i\omega(t-\boldsymbol{u}\cdot\boldsymbol{r})\},$$

where $\boldsymbol{A}$, $\boldsymbol{C}$ and $\boldsymbol{u}$ are constant vectors. Prove that

$$|\boldsymbol{u}|^{-1} = (k\varepsilon_0\mu_0)^{1/2}, \quad \varepsilon_0 k\boldsymbol{E} = \boldsymbol{H}\times\boldsymbol{u}, \quad \mu_0\boldsymbol{H} = \boldsymbol{u}\times\boldsymbol{E}.$$

If such a train of waves is incident normally on the plane face of a semi-infinite block of glass of refractive index n, show that the intensities of the incident, reflected and refracted waves are in the ratio $(n+1)^2 : (n-1)^2 : 4n$.

4. A plane electromagnetic wave in free space is incident at an angle θ on the plane boundary of a non-conducting medium of dielectric constant K and of permeability μ_0, the wave being polarized with the electric vector in the plane of incidence. Prove that the angle of refraction ϕ is given by $\sin\theta = \sqrt{K}\sin\phi$ and that the ratio of the electric field strength of the reflected wave to that of the incident wave is given by

$$\tan(\theta-\phi)/\tan(\theta+\phi).$$

5. If $\boldsymbol{l}$, $\boldsymbol{m}$, $\boldsymbol{n}$ are a right-handed triad of constant unit vectors, show that the fields

$$\boldsymbol{E} = \frac{A}{\sqrt{\varepsilon}} \exp\left\{i\omega\left(t - \frac{\boldsymbol{n}\cdot\boldsymbol{r}}{v}\right)\right\}\boldsymbol{l}, \quad \boldsymbol{B} = A\sqrt{\mu} \exp\left\{i\omega\left(t - \frac{\boldsymbol{n}\cdot\boldsymbol{r}}{v}\right)\right\}\boldsymbol{m},$$

with $v = (\mu\varepsilon)^{-1/2}$, satisfy Maxwell's equations for a uniform non-conducting medium of permittivity ε and permeability μ, and that they form a plane-polarized plane wave travelling with velocity v in the direction $\boldsymbol{n}$.

Such a wave travelling in the vacuum is incident on the plane face $z = 0$ of a semi-infinite uniform dielectric medium of permittivity ε and permeability μ_0. The electric vector is in the plane of incidence, and the acute angle between the direction of propagation and the z-axis is $\theta_1 = \tan^{-1}\sqrt{(\varepsilon/\varepsilon_0)}$. Show that the boundary conditions at the surface are satisfied if there is no reflected wave, if the transmitted wave has the same plane of polarization as the incident, and if it is propagated in the plane of incidence in a direction making an angle $\theta = \frac{1}{2}\pi - \theta_1$ with the normal.

11.5 Total internal reflection

Total internal reflection is a phenomenon well known in elementary physics. It occurs when a ray, travelling through an (optically) dense medium, meets the boundary surface at an angle of incidence which is greater than the *critical angle.* The critical angle is that angle of incidence (in the dense medium) which corresponds to an angle of refraction (in the external medium, of a right angle (see Fig. 11.8). The difference between the situations of Figs. 11.7 and 11.8 is that in Fig. 11.8 the incident and reflected rays, i.e. the region with $x > 0$, are in a medium with permittivity $K\varepsilon_0$, and their velocity of propagation is $v = 1/\sqrt{(K\varepsilon_0\mu_0)}$; the region $x < 0$, containing the refracted ray, is a medium with permittivity ε_0 corresponding to a velocity of propagation $c = 1/\sqrt{(\varepsilon_0\mu_0)}$. The results obtained in § 11.3 can therefore be adapted to the present case by interchanging the roles of c, v wherever they occur in the formulae.

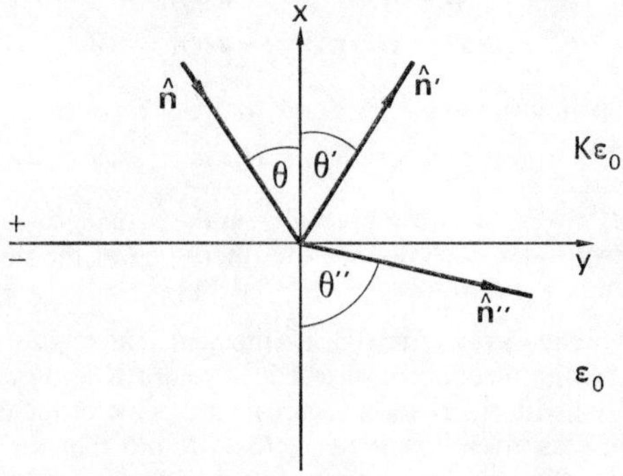

FIG. 11.8

So long as the angle of incidence θ is less than the critical angle, the analysis of § 11.3 applies without change except for replacing c by v, and vice versa. We are here concerned with the case in which θ exceeds the critical value. We shall see that this leads to the appearance of imaginary, or complex, values instead of some of the real values in § 11.3, and we give here a changed interpretation which takes account of this appearance of complex quantities in the analysis.

The boundary conditions which we applied in § 11.3 must continue to hold even for angles of incidence exceeding the critical angle, so we deduce, because of the interchange of v and c, that

$$v = \omega/k, \quad c = \omega/k'' = v\sqrt{K}. \tag{11.85}$$

As before the frequency ω must be common to all three rays, and they must lie in one plane, i.e. $\hat{\boldsymbol{n}}\cdot\boldsymbol{k} = \hat{\boldsymbol{n}}'\cdot\boldsymbol{k} = \hat{\boldsymbol{n}}''\cdot\boldsymbol{k} = 0$ [eqn. (11.41)] for the same reasons. The condition which led to Snell's law (11.43) now gives

$$\hat{\boldsymbol{n}}''\cdot\boldsymbol{j} = (k/k'')(\hat{\boldsymbol{n}}\cdot\boldsymbol{j}) = (k/k'')\sin\theta = (c/v)\sin\theta = (\sqrt{K})\sin\theta.$$

This must hold for *all* possible angles of incidence including angles of incidence greater than the critical angle. We write

$$\hat{\boldsymbol{n}}''\cdot\boldsymbol{j} = \cosh\phi, \quad \frac{c}{\cosh\phi} = \frac{v}{\sin\theta}$$

when $\sin\theta > v/c = 1/\sqrt{K}$. Since $\hat{\boldsymbol{n}}''\cdot\boldsymbol{k} = 0$ and $\hat{\boldsymbol{n}}''$ is a unit vector, we must also have

$$\cos\theta'' = \hat{\boldsymbol{n}}''\cdot\boldsymbol{i} = \sqrt{(1-\cosh^2\phi)} = \mathrm{i}\sinh\phi.$$

Therefore

$$\hat{\boldsymbol{n}}'' = \{\mathrm{i}\sinh\phi \quad \cosh\phi \quad 0\}. \tag{11.86}$$

The case of critical incidence therefore corresponds to $\phi = 0$. The application of the boundary conditions now leads to the following results [instead of those of eqn. (11.65)]

$$b_p' = \frac{v\cos\theta'' - c\cos\theta}{v\cos\theta'' + c\cos\theta}\, b_p = \frac{\mathrm{i}\sin\theta\sinh\phi - \cos\theta\cosh\phi}{\mathrm{i}\sin\theta\sinh\phi + \cos\theta\cosh\phi}\, b_p$$

$$= -\frac{\cos(\theta + \mathrm{i}\phi)}{\cos(\theta - \mathrm{i}\phi)}\, b_p = \mathrm{e}^{(\mathrm{i}\pi + 2\alpha)} b_p, \tag{11.87}$$

where $\tan\alpha = \tan\theta\tanh\phi$. Similarly,

$$b_n' = \frac{v\cos\theta - c\cos\theta''}{v\cos\theta + c\cos\theta''}\, b_n = \frac{\sin\theta\cos\theta - \mathrm{i}\cosh\phi\sinh\phi}{\sin\theta\cos\theta + \mathrm{i}\cosh\phi\sinh\phi}\, b_n = \mathrm{e}^{-2\mathrm{i}\beta} b_n, \tag{11.88}$$

where
$$\tan\beta = \frac{\cosh\phi\sinh\phi}{\cos\theta\sin\theta}.$$

This shows that $|b_p'| = |b_p|$, $|b_n'| = |b_n|$ and that the reflected ray differs from the incident ray in phase, but not in amplitude.

The results corresponding to (11.65–6) for the refracted ray are

$$b_p'' = \frac{2v\cos\theta}{v\cos\theta''+c\cos\theta}\,b_p = \frac{2\sin\theta\cos\theta}{\cos\theta\cosh\phi+\mathrm{i}\sin\theta\sinh\phi}\,b_p$$
$$= (P^{-1}\sin 2\theta)\mathrm{e}^{-\mathrm{i}\alpha}b_p, \tag{11.89}$$

where $P = |\cos(\theta-\mathrm{i}\phi)|$;

$$b_n'' = \frac{2v\cos\theta}{v\cos\theta+c\cos\theta''}\,b_n = \frac{2\sin\theta\cos\theta}{\sin\theta\cos\theta+\mathrm{i}\sinh\phi\cosh\phi}\,b_n$$
$$= (N^{-1}\sin 2\theta)\mathrm{e}^{-\mathrm{i}\beta}b_n, \tag{11.90}$$

where $N = |\sin\theta\cos\theta+\mathrm{i}\sinh\phi\cosh\phi|$.

We now consider the interpretation of the expressions we have obtained, and the effect of the complex components. Since the quantities referring to the incident and reflected rays are real, as in § 11.3, there is no change to be made to the interpretation for the region $x > 0$. We look at the region $x < 0$ where the field is given by $\boldsymbol{B}''$, $\boldsymbol{E}''$.

The exponential factor in $\boldsymbol{B}''$, $\boldsymbol{E}''$, when the angle of incidence exceeds the critical angle, is given by

$$\exp\mathrm{i}\{\omega t-k''(\hat{\boldsymbol{n}}''\cdot\boldsymbol{r})\} = \exp\mathrm{i}\omega\{t-\mathrm{i}(x/c)\sinh\phi-(y/c)\cosh\phi\}$$
$$= \mathrm{e}^{\omega(x/c)\sinh\phi}\exp\mathrm{i}\omega\{t-(y/c)\cosh\phi\}.$$
$$\therefore\quad \boldsymbol{B}'' = (\boldsymbol{b}''\mathrm{e}^{\omega(x/e)\sinh\phi})\exp\mathrm{i}\omega\{t-(y/c)\cosh\phi\}, \tag{11.91}$$
$$\boldsymbol{E}'' = c(\boldsymbol{b}''\times\hat{\boldsymbol{n}}'')\mathrm{e}^{\omega(x/c)\sinh\phi}\exp\mathrm{i}\omega\{t-(y/c)\cosh\phi\}. \tag{11.92}$$

Since the factor $\mathrm{e}^{\omega(x/c)\sinh\phi}$ is real we interpret this as harmonic waves, $\exp\mathrm{i}\omega\{t-(y/c)\cosh\phi\}$, with amplitudes which depend on x through the factor $\mathrm{e}^{\omega(x/c)\sinh\phi}$. The harmonic wave is propagated in the y-direction, i.e. along the surface of separation, with speed

$$\frac{c}{\cosh\phi} = \frac{v}{\sin\theta}.$$

Because the region is one for which $x < 0$ the dependence of the amplitude of the wave on distance from the surface is a rapid exponential decrease; the rate of decrease is greater for higher frequencies. The reader can picture the situation by visualizing a train of straight (water) waves approaching a

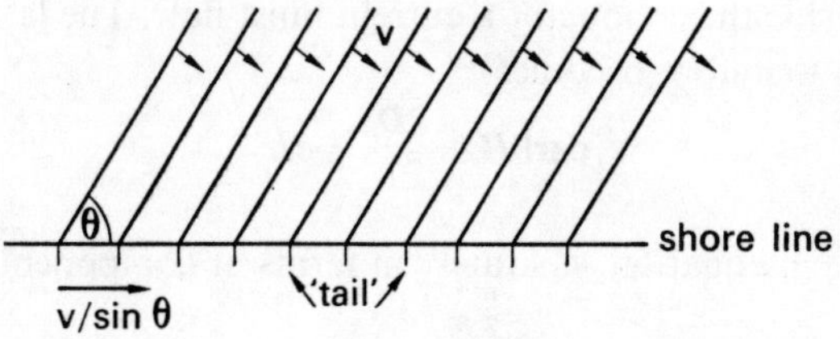

FIG. 11.9

straight shore obliquely, Fig. 11.9. The "tail" of each wave dies out as it runs a short distance up the shore, and each tail runs directly along the shore line as the wave arrives; this is the velocity $v/\sin\theta$. The fact that $\boldsymbol{B}''$, $\boldsymbol{E}''$ are not zero but decrease exponentially for $x < 0$ shows that there is slight penetration by the field into the "forbidden" region.

The results obtained in § 11.4 can be used to investigate the transport of energy by the fields we are considering. The situation in the region $x > 0$, except for the interchange of v and c, is exactly as given in § 11.3 and the transport of energy is given by eqns. (11.76–8). Since $|b_p'| = |b_p|$ and $|b_n'| = |b_n|$, it follows from (11.79) that there is no net flow of energy, either toward or away from, the surface, in the region $x > 0$. For the region $x < 0$ we use (11.80) and obtain

$$-\boldsymbol{P}_{-}\cdot\boldsymbol{i} = \mathrm{i}c \sinh\phi(|b_p''|^2 + |b_n''|^2)/(2\mu_0), \qquad (11.93)$$

since $\cos\theta'' = \mathrm{i}\sinh\phi$. Since this is a purely imaginary quantity, it implies that there is *no* transport of energy by the field $\boldsymbol{E}''$, $\boldsymbol{B}''$. Therefore the whole of the energy brought to the surface by the incident ray is carried back into the medium by the reflected ray. Hence there is total internal reflection.

11.6 Propagation of waves in a conducting medium

In this section we consider the effect of conductivity when an electromagnetic field is established in a conducting medium. We assume a uniform medium specified by the constants: permittivity ε, permeability μ, conductivity σ; and we assume that these constants are independent of position, time and field strength. The equations to be satisfied now are

$$\operatorname{div}\boldsymbol{D} = \varrho, \quad \operatorname{div}\boldsymbol{B} = 0, \quad \mathbf{curl}\,\boldsymbol{E} + \frac{\partial\boldsymbol{B}}{\partial t} = \boldsymbol{0}, \quad \mathbf{curl}\,\boldsymbol{H} - \frac{\partial\boldsymbol{D}}{\partial t} = \boldsymbol{J},$$

where
$$\boldsymbol{D} = \varepsilon\boldsymbol{E}, \quad \boldsymbol{H} = \boldsymbol{B}/\mu, \quad \boldsymbol{J} = \sigma\boldsymbol{E}.$$

As in § 11.1 we look for a field which depends only on x, t and assume that $\varrho = 0$. We cannot assume that $\boldsymbol{J} = \boldsymbol{0}$, as we did in § 11.1, for if there is an electric field inside the conductor a current must flow. The last of Maxwell's field equations therefore becomes

$$\mathbf{curl}\,\boldsymbol{H}-\frac{\partial \boldsymbol{D}}{\partial t}=\sigma \boldsymbol{E}.$$

We consider each equation separately in terms of components.

$$\operatorname{div}\boldsymbol{D}=\operatorname{div}(\varepsilon\boldsymbol{E})=0: \qquad \frac{\partial E_x}{\partial x}=0. \tag{11.94}$$

$$\operatorname{div}\boldsymbol{B}=0: \qquad \frac{\partial B_x}{\partial x}=0. \tag{11.95}$$

$$\mathbf{curl}\,\boldsymbol{E}+\frac{\partial \boldsymbol{B}}{\partial t}=\boldsymbol{0}: \qquad \frac{\partial B_x}{\partial t}=0; \tag{11.96}$$

$$-\frac{\partial E_z}{\partial x}+\frac{\partial B_y}{\partial t}=0, \tag{11.97}$$

$$\frac{\partial E_y}{\partial x}+\frac{\partial B_z}{\partial t}=0. \tag{11.98}$$

$$\mathbf{curl}\,\boldsymbol{H}-\frac{\partial \boldsymbol{D}}{\partial t}=\sigma\boldsymbol{E}; \qquad -\varepsilon\frac{\partial E_x}{\partial t}=\sigma E_x, \tag{11.99}$$

$$-\frac{1}{\mu}\frac{\partial B_z}{\partial x}-\varepsilon\frac{\partial E_y}{\partial t}=\sigma E_y, \tag{11.100}$$

$$\frac{1}{\mu}\frac{\partial B_y}{\partial x}-\varepsilon\frac{\partial E_z}{\partial t}=\sigma E_z. \tag{11.101}$$

From (11.95–6) we can take $B_x = 0$ as before. However, (11.94) implies that E_x can depend only on t, and (11.99) gives this dependence as $E_x = E_{x0}e^{-\sigma t/\varepsilon}$. This shows that the longitudinal component of $\boldsymbol{E}$ dies out exponentially with t, but is not identically zero as is B_x. However, the time (of relaxation) is very short, i.e. E_x dies out very rapidly so that we can effectively consider this as zero also. (The term "longitudinal" is used because E_x is the component of $\boldsymbol{E}$ in the direction of propagation.)

The remaining equations fall into two independent pairs (11.97) and (11.101) involving E_z, B_y, and (11.98) and (11.100) involving E_y, B_z. It is easy to see that all four of these components satisfy the scalar equation

$$\frac{\partial^2 f}{\partial x^2}=\mu\varepsilon\frac{\partial^2 f}{\partial t^2}+\mu\sigma\frac{\partial f}{\partial t}. \tag{11.102}$$

This differs from the wave equation in containing the extra term $\mu\sigma\,\partial f/\partial t$; it is this term which gives rise to the special properties of the conducting medium. If the medium were non-conducting ($\sigma = 0$), the velocity v of propagation of waves would be given by $\mu\varepsilon = 1/v^2$.

We will, for simplicity, consider one of the independent pairs of equations, viz. (11.98) and (11.100), and seek a solution corresponding to (11.12–13), a harmonic wave. We notice that the vectors $\boldsymbol{B}, \boldsymbol{E}$ are perpendicular to each other and to the direction of propagation. We write

$$E_y = E \exp \mathrm{i}(\omega t - kx), \quad B_z = B \exp \mathrm{i}(\omega t - kx), \tag{11.103}$$

where E, B are constants. From (11.98) we obtain

$$-\mathrm{i}kE \exp \mathrm{i}(\omega t - kx) = -\mathrm{i}\omega B \exp \mathrm{i}(\omega t - kx),$$

and from (11.100)

$$\mathrm{i}kB \exp \mathrm{i}(\omega t - kx) + \mu\varepsilon(-\mathrm{i}k)E \exp \mathrm{i}(\omega t - kx) = \sigma E \exp \mathrm{i}(\omega t - kx).$$

These two last equations reduce to

$$kE = \omega B, \tag{11.104}$$

and

$$\mathrm{i}kB = (\mathrm{i}\omega\varepsilon\mu + \mu\sigma)E. \tag{11.105}$$

Therefore

$$k^2 = \omega^2\varepsilon\mu - \mathrm{i}\mu\sigma\omega. \tag{11.106}$$

A convenient, quick method to obtain the results for a conducting medium is to replace ε for the medium by $\varepsilon - \mathrm{i}\sigma/\omega$. Thus, for the refracted ray, $k''^2 = \omega^2/v^2 = \omega^2\varepsilon\mu$ in eqn. (11.22), and $k''^2 = \omega^2\mu(\varepsilon - \mathrm{i}\sigma/\omega)$ in eqn. (11.106).

This implies that k is a complex number, and that E, B differ in their arguments as well as moduli. (In § 11.1 both amplitudes were real—or had the same complex arguments.) We write

$$k = \alpha - \mathrm{i}\beta \qquad (\alpha, \beta \text{ real}) \text{ so that,}$$

$$\alpha^2 - \beta^2 = \omega^2\varepsilon\mu, \quad 2\alpha\beta = \mu\sigma\omega, \tag{11.107}$$

and obtain for the field strengths

$$\begin{aligned} E_y &= E\mathrm{e}^{-\beta x} \exp \mathrm{i}(\omega t - \alpha x), \\ B_z &= B\mathrm{e}^{-\beta x} \exp \mathrm{i}(\omega t - \alpha x). \end{aligned} \tag{11.108}$$

These represent harmonic waves having a phase velocity ω/α, propagated in the positive x-direction, but having amplitudes $E\mathrm{e}^{-\beta x}$, $B\mathrm{e}^{-\beta x}$ which decrease with increasing x, i.e. the wave is attenuated as it is propagated through the conductor. Both the attenuation factor $\mathrm{e}^{-\beta x}$ and the phase velocity ω/α vary with the frequency ω. This dependence of the velocity of propagation on ω is a new feature; the different frequencies travel with different speeds. If a wave containing more than one frequency enters the conductor its profile and energy distribution alter as it travels. In general this means that energy is propagated with a velocity different from the phase-velocity and the concept of "group-velocity" arises. This dependence of velocity on frequency causes "dispersion". However, the dominant effect of conductivity is that of attenuation, or absorption, of the wave in the conductor. Because of this a wave never penetrates very far into a conducting medium, the higher the frequency, or the greater the conductivity, the less the penetration. This results in the "skin effect" which means that all oscillating fields are confined to a narrow layer, or skin, near the surface of the conductor. For a perfect conductor ($\sigma \to \infty$) the penetration is zero and there is no field at all inside a perfect conductor.

In order to see more clearly the effect of the conductivity we obtain expressions for α, β corresponding to small and large (but not infinite) values of σ.

When σ is small

$$k = (\omega^2/v^2 - \mathrm{i}\mu\sigma\omega)^{1/2} = (\omega/v)\{1-\mathrm{i}\sigma/(\varepsilon\omega)\}^{1/2}$$
$$\approx (\omega/v)\{1-\mathrm{i}\sigma(2\varepsilon\omega)\}.$$

Therefore $$\alpha = \omega/v, \qquad \beta = \sigma/(2\varepsilon v), \tag{11.109}$$

where we have written $v = (\varepsilon\mu)^{-1/2}$, v being the velocity of propagation of a wave in a similar medium without conductivity. In this case, σ small, the velocity of propagation α/ω has the "normal" value v, and the attenuation is slight.

When σ is large (compared with $\varepsilon\omega$), as in "good conductors" such as copper, we write from (11.106),

$$k^2 = -\mathrm{i}\mu\sigma\omega; \qquad k = (1-\mathrm{i})(\mu\sigma\omega/2)^{1/2}.$$

Therefore $$\alpha = (\mu\sigma\omega/2)^{1/2} = \beta. \tag{11.110}$$

In this case the velocity of propagation is $\omega/\alpha = v(2\omega\varepsilon/\sigma)^{1/2}$. This shows that the velocity of propagation is reduced well below the "normal" value v, and the attenuation is heavy.

The extent of the penetration is measured by the *skin-depth* δ. This is the distance in which the amplitude of the wave is reduced to $1/\mathrm{e}$ of its value originally. Therefore

$$\delta = \frac{1}{\beta} \approx \left(\frac{2}{\mu\sigma\omega}\right)^{1/2} \tag{11.111}$$

for large values of σ, or high frequencies. These are the cases when the effect is important.

At certain levels in the earth's atmosphere the gas is ionized to an appreciable extent so that the atmosphere at this level has an appreciable conductivity. Because of this fact radio waves cannot penetrate the layer, but are reflected back to the surface of the earth. Roughly speaking, this explains why reception of radio signals is possible although the radio transmitter is well "out of sight" of the receiver and the rays have had to come round the curvature of the earth's surface. They have been reflected around the curvature from one of these conducting layers (e.g. the Appleton layer).

This process of reflection by a conductor can be investigated by a discussion similar to that used in § 11.3. We consider a surface $z = 0$ separating a region $z > 0$, which is a vacuum with permittivity and permeability ε_0, μ_0, from the conducting region $z < 0$ where the permittivity, permeability and conductivity have the respective, uniform, values ε, μ, σ. We assume a plane wave incident on the surface together with a reflected and a refracted wave and the boundary conditions, corresponding to (11.24–27), are now

$$\boldsymbol{i}\cdot(\boldsymbol{B}_+ - \boldsymbol{B}_-) = \boldsymbol{i}\cdot(\boldsymbol{B}+\boldsymbol{B}'-\boldsymbol{B}'') = 0, \tag{11.112}$$

$$\boldsymbol{i}\cdot(\varepsilon_0\boldsymbol{E}_+ - \varepsilon\boldsymbol{E}_-) = \boldsymbol{i}\cdot(\varepsilon_0\boldsymbol{E}+\varepsilon_0\boldsymbol{E}'-\varepsilon\boldsymbol{E}'') = 0, \tag{11.113}$$

$$\boldsymbol{i}\times(\boldsymbol{H}_+ - \boldsymbol{H}_-) = \boldsymbol{i}\times(\boldsymbol{B}/\mu_0+\boldsymbol{B}'/\mu_0-\boldsymbol{B}''/\mu) = \boldsymbol{0}, \tag{11.114}$$

$$\boldsymbol{i}\times(\boldsymbol{E}_+ - \boldsymbol{E}_-) = \boldsymbol{i}\times(\boldsymbol{E}+\boldsymbol{E}'-\boldsymbol{E}'') = \boldsymbol{0}. \tag{11.115}$$

We adopt the notation and representations used in § 11.3 [see Fig. 11.7 and eqns. (11.36) to (11.39)]. First, the requirement that the exponential factors are identical for all points and times on the boundary surfaces leads to

$$\omega = \omega' = \omega'', \quad k = \omega/c = k'.$$

But for the refracted ray complex values must be used for k'', $\cos\theta''$, $\sin\theta''$ though the forms of the results remain the same. Since $k''^2 = \omega^2\varepsilon\mu - \mathrm{i}\mu\sigma\omega$ [eqn. (11.106)] and

$$k\sin\theta = k\sin\theta' = k''\sin\theta'', \quad k\cos\theta = k\cos\theta' = k''\cos\theta'',$$

both $\sin\theta''$ and $\cos\theta''$ are complex. We have

$$\begin{aligned} k''\cos\theta'' &= (k''^2 - k''^2\sin^2\theta'')^{1/2} \\ &= (\omega^2\mu\varepsilon - \mathrm{i}\omega\sigma\mu - k^2\sin^2\theta)^{1/2}, \end{aligned}$$

where $\mu\varepsilon = 1/v^2$, $k^2 = \omega^2/c^2$.

Therefore $$k''\cos\theta'' = p - \mathrm{i}q, \tag{11.116}$$

where $$p^2 - q^2 = (\omega^2/v^2) - (\omega^2/c^2)\sin^2\theta \qquad (> 0) \tag{11.117}$$

$$2pq = \omega\sigma\mu, \tag{11.118}$$

and we may take $p, q > 0$.

Now, since $\hat{\boldsymbol{n}}'' = \{-\cos\theta'' \quad \sin\theta'' \quad 0\}$ the exponential factor is

$$\begin{aligned} \exp i\{\omega t - k''(\hat{\boldsymbol{n}}''\cdot\boldsymbol{r})\} &= \exp i\{\omega t + k''x\cos\theta'' - k''y\sin\theta''\} \\ &= e^{qx}\exp i\{\omega t + px - yk\sin\theta\}. \end{aligned} \tag{11.119}$$

This is a harmonic wave in which surfaces of constant amplitude, x = constant, are different from surfaces of constant phase, $px - ky\sin\theta$ = constant. Since $q > 0$ we see that as the wave penetrates further into the metal ($x < 0$) the amplitude must decrease, and the "skin effect" is displayed as before. The planes of constant phase advance in a direction given by the unit vector $\{-\cos\psi \quad \sin\theta \quad 0\}$, i.e. ψ is the effective angle of refraction. The velocity of propagation is given by

$$u = \omega/(p^2+k^2\sin^2\theta)^{1/2} = \omega/(q^2+\omega^2/v^2)^{1/2}. \tag{11.120}$$

The direction ψ is given by

$$\cos\psi = p/u, \quad \sin\psi = k\sin\theta/u = (k/\omega)(q^2+\omega^2/v^2)^{1/2}\sin\theta. \tag{11.121}$$

The last result shows that the effective refractive index is

$$n = \frac{\sin\theta}{\sin\psi} = \frac{k}{\omega(q^2+\omega^2/v^2)^{1/2}}$$

and this depends upon the angle of incidence θ, through the dependence of q on θ given by eqns. (11.117–118). This is yet a further complication in the effect of the metallic surface on the wave. When we look at the Fresnel formulae in (11.65–66) the occurrence of the complex quantities in $\cos\theta''$ shows that the process of reflection at a metallic surface is by no means so simple as it appears. We can deduce, in general, that unlike reflection from a non-conducting surface, there is a difference of phase between b_p', b_p and b_n', b_n; also that $|b_n'| = |b_n|$ and $|b_p'| = |b_p|$. This latter result implies that all the energy in the incident ray is reflected by the surface. The detailed discussion of the relations between the amplitudes of the various rays is given in advanced books on electromagnetic theory; the calculations are tedious but mostly elementary.

Exercises 11.6

1. State the electromagnetic field equations, and deduce that for an uncharged medium of homogeneous properties ε, μ, σ, the electric field satisfies the equation

$$\nabla^2\boldsymbol{E} = \varepsilon\mu\frac{\partial^2\boldsymbol{E}}{\partial t^2} + \sigma\mu\frac{\partial\boldsymbol{E}}{\partial t},$$

where a system of rectangular cartesian coordinates and components is implied.

The plane $z = 0$ is the interface between two media; for $z < 0$ the medium is of zero conductivity, and for $z > 0$ the conductivity of the medium is σ. The dielectric constants and permeabilities are ε_1, μ_1 and ε_2, μ_2 respectively. An electromagnetic wave, for which

$$\boldsymbol{E} = E_0\{0 \quad \exp[i\omega(t - z/v_1)] \quad 0\}$$

is incident on the interface from $z < 0$. State the value of v_1, and show that the transmitted wave varies as

$$\exp\{i\omega(t - kz)\},$$

where
$$k^2 = \varepsilon_2\mu_2 - \frac{i\sigma\mu_2}{\omega}.$$

Determine the complete electromagnetic field, and find the ratio of the amplitudes of the reflected and incident electric fields.

2. The electric intensity $\boldsymbol{E}$ of a wave in a medium of conductivity σ, permeability μ and permittivity ε is given by

$$\boldsymbol{E} = \boldsymbol{i}e^{i(\omega t - kz)}$$

where $\boldsymbol{i}$ is a unit vector along the x-axis. Determine k and the magnetic field.

A linearly polarized plane wave of period $2\pi/\omega$ is incident normally from free space on the plane face of a semi-infinite metal. If $\sigma \gg \omega\varepsilon$ and $\sigma \gg \omega\mu$ show that the amplitude of the electric vector is reduced on reflection in the ratio $1 - \sqrt{(\omega\mu/2\pi\sigma)}$ approximately.

3. A uniform isotropic material has conductivity σ, permeability μ and permittivity ε. Show that Maxwell's equations have a solution in which the only non-zero components of $\boldsymbol{E}$ and $\boldsymbol{H}$ are E_y and H_z, and in which $E_y = Ae^{-\omega mx/c} \cos \omega(t - nx/c)$ where $n^2 - m^2 = c^2\varepsilon\mu$ and $nm = \frac{1}{2}c^2\mu\sigma/\omega$.

An infinite slab of this material fills the space $x \geqslant 0$ and a plane electromagnetic wave with $\boldsymbol{E} = \{0 \; \mathcal{E} \cos \omega(t - x/c) \; 0\}$ is incident on the slab. Determine the reflected and transmitted waves.

4. Prove that in a medium of conductivity σ, permeability μ and permittivity ε there is a solution of Maxwell's equations of the form

$$\boldsymbol{E} = \boldsymbol{i}E_0 \exp(i\omega t - in\omega z/c - \alpha\omega z/c),$$
$$\boldsymbol{B} = \boldsymbol{j}B_0 \exp(i\omega t - in\omega z/c - \alpha\omega z/c),$$

where $B_0 = (n - i\alpha)E_0/c^2$ and $(n - i\alpha)^2 = c^2(\mu\varepsilon - i\sigma\mu/\omega)$.

Material of conductivity σ fills the region $0 \leqslant z$, and a plane harmonic wave in vacuo of frequency $\omega/2\pi$, given by

$$\boldsymbol{E} = \boldsymbol{i}A \exp(i\omega t - i\omega z/c), \qquad \boldsymbol{B} = \boldsymbol{j}(A/c) \exp(i\omega t - i\omega z/c)$$

in the region $z \leqslant 0$, is incident on the face $z = 0$. Prove that the ratio of the energy of the wave reflected by the material to the energy of the incident wave is $(n-1)/(n+1)$. Find the fraction of the incident energy which crosses a plane distant z (>0) from the surface of separation.

11.7 Waveguides

A waveguide is a hollow metal tube, made of highly conducting material down which electromagnetic waves are propagated. In this section we investigate the effects of the conducting walls on the waves which can pass along the tube. In our theoretical discussion we assume that the tube is a cylinder whose generators are parallel to the z-axis and that the medium inside the tube is a vacuum (ε_0, μ_0), the walls of the tube being perfect conductors. Waveguides used in practice approximate to this, but have imperfectly conducting walls, may have bends or junctions in the tube, and may even be partly occupied by dielectric media. We do not discuss the effects of these latter modifications, but merely establish the basic types of wave that can travel in a straight waveguide.

The discussion has two aspects: first, we look for a special form of solution of Maxwell's equations which corresponds to a wave propagated in the z-direction, but whose amplitude is not uniform across the tube. (This is the chief difference between our analysis and our previous discussion of plane waves.) Second, we see what restrictions the presence of boundary walls of perfect conductors place on the solutions so obtained. We use complex exponentials to represent the oscillating fields and assume no charge or currents inside the tube. (There will, of course, be charges and currents on the walls of the tube to correspond with the fields.)

We seek solutions of the form

$$\boldsymbol{E} = \boldsymbol{e}(x, y) \exp\{\mathrm{i}(\omega t - \gamma z)\}, \quad \boldsymbol{B} = \boldsymbol{b}(x, y) \exp\{\mathrm{i}(\omega t - \gamma z)\}. \tag{11.122}$$

Here, unlike previous discussions, $\boldsymbol{e}$ and $\boldsymbol{b}$ depend upon x, y, but z only occurs in the exponential factor. These expressions correspond to harmonically varying fields in which the planes z = constant are planes of constant phase; these planes advance with the phase-velocity ω/γ. We consider, when necessary, complex values for γ which will correspond to attenuation of the wave. But, so long as γ is real (and positive) the solution corresponds to propagation without attenuation.

We substitute the expressions (11.122) into Maxwell's equations and obtain

$$\operatorname{div} \boldsymbol{E} = 0: \quad \operatorname{div} \boldsymbol{e} - \mathrm{i}\gamma(\boldsymbol{k}\cdot\boldsymbol{e}) = 0; \quad \frac{\partial e_x}{\partial x} + \frac{\partial e_y}{\partial y} - i\gamma e_z = 0; \tag{11.123}$$

$$\operatorname{div} \boldsymbol{B} = 0: \quad \operatorname{div} \boldsymbol{b} - \mathrm{i}\gamma(\boldsymbol{k}\cdot\boldsymbol{b}) = 0; \quad \frac{\partial b_x}{\partial x} + \frac{\partial b_y}{\partial y} - \mathrm{i}\gamma b_z = 0; \tag{11.124}$$

$$\mathbf{curl}\, \boldsymbol{E} + \frac{\partial \boldsymbol{B}}{\partial t} = \boldsymbol{0}: \quad \mathbf{curl}\, \boldsymbol{e} - \mathrm{i}\gamma(\boldsymbol{k}\times\boldsymbol{e}) + \mathrm{i}\omega\boldsymbol{b} = \boldsymbol{0}, \tag{11.125}$$

or

$$\frac{\partial e_z}{\partial y} + \mathrm{i}\gamma e_y + \mathrm{i}\omega b_x = 0, \tag{11.125a}$$

$$-\frac{\partial e_z}{\partial x} - i\gamma e_x + i\omega b_y = 0, \qquad (11.125b)$$

$$\frac{\partial e_y}{\partial x} - \frac{\partial e_x}{\partial y} + i\omega b_z = 0; \qquad (11.125c)$$

$$\mathbf{curl}\, \boldsymbol{H} - \frac{\partial \boldsymbol{D}}{\partial t} = \boldsymbol{0}; \qquad \mathbf{curl}\, \boldsymbol{b} - i\gamma(\boldsymbol{k}\times\boldsymbol{b}) - i(\omega/c^2)\boldsymbol{e} = \boldsymbol{0}, \qquad (11.126)$$

or

$$\frac{\partial b_z}{\partial y} + i\gamma b_y - \frac{i\omega}{c^2} e_x = 0, \qquad (11.126a)$$

$$-\frac{\partial b_z}{\partial x} - i\gamma b_x - \frac{i\omega}{c^2} e_y = 0, \qquad (11.126b)$$

$$\frac{\partial b_y}{\partial x} - \frac{\partial b_x}{\partial y} - \frac{i\omega}{c^2} e_z = 0. \qquad (11.126c)$$

In eqn. (11.126) we have used the relation $\varepsilon_0\mu_0 = 1/c^2$. From (11.125) and (11.126) we see that

$$\boldsymbol{k}\cdot\mathbf{curl}\, \boldsymbol{e} + i\omega(\boldsymbol{b}\cdot\boldsymbol{k}) = 0, \quad \boldsymbol{k}\cdot\mathbf{curl}\, \boldsymbol{b} - i(\omega/c^2)(\boldsymbol{e}\cdot\boldsymbol{k}) = 0. \qquad (11.127)$$

This shows that there are three possible types of solution.

1. One for which $\boldsymbol{k}\cdot\mathbf{curl}\, \boldsymbol{e} = 0 = (\boldsymbol{b}\cdot\boldsymbol{k})$ in which case there is a *transverse magnetic* field; these are called TM-waves (sometimes E-waves).
2. One for which $\boldsymbol{k}\cdot\mathbf{curl}\, \boldsymbol{b} = 0 = (\boldsymbol{e}\cdot\boldsymbol{k})$ in which case there is a *transverse electric* field; these are called TE-waves (sometimes H-waves).
3. One for which both $\boldsymbol{k}\cdot\mathbf{curl}\, \boldsymbol{e}$ and $\boldsymbol{k}\cdot\mathbf{curl}\, \boldsymbol{b}$ vanish, in which case both electric and magnetic vectors are transverse. These are called TEM-waves. The general solution of the form (11.122) is an arbitrary linear combination of these three types.

We continue the discussion further for TM-waves, and give, without derivation, the corresponding results for TE-waves.

TM-waves

For these waves $b_z = 0$, and the equations show that all other components of $\boldsymbol{b}$ and $\boldsymbol{e}$ can be expressed in terms of e_z. Since $b_z = 0$ everywhere,

$$\frac{\partial b_x}{\partial x} + \frac{\partial b_y}{\partial y} = 0, \qquad \frac{\partial e_x}{\partial x} + \frac{\partial e_y}{\partial y} - i\gamma e_z = 0.$$

$$b_y = \left(\frac{\omega}{\gamma c^2}\right) e_x, \qquad \frac{\partial e_z}{\partial y} + i\gamma\left(1 - \frac{\omega^2}{\gamma^2 c^2}\right) e_y = 0,$$

$$b_x = -\left(\frac{\omega}{\gamma c^2}\right)e_y, \qquad \frac{\partial e_z}{\partial x} + i\gamma\left(1-\frac{\omega^2}{\gamma^2 c^2}\right)e_x = 0,$$

$$\frac{\partial b_y}{\partial x} - \frac{\partial b_x}{\partial y} = \frac{i\omega}{c^2}e_z, \qquad \frac{\partial e_y}{\partial x} - \frac{\partial e_x}{\partial y} = 0.$$

Elimination of all components except e_z leads now to

$$\frac{\partial^2 e_z}{\partial x^2} + \frac{\partial^2 e_z}{\partial y^2} = i\gamma\left(1-\frac{\omega^2}{\gamma^2 c^2}\right)\left(\frac{\partial e_x}{\partial x} + \frac{\partial e_y}{\partial y}\right) = -\left(\gamma^2 - \frac{\omega^2}{c^2}\right)e_z = -\nu^2 e_z.$$

Hence we may write

$$\frac{\partial^2 \phi}{\partial x^2} + \frac{\partial^2 \phi}{\partial y^2} + \nu^2 \phi = 0, \tag{11.128}$$

$$e_x^{\text{TM}} = -\frac{i\gamma}{\nu^2}\frac{\partial \phi}{\partial x}, \quad e_y^{\text{TM}} = -\frac{i\gamma}{\nu^2}\frac{\partial \phi}{\partial y}, \quad e_z^{\text{TM}} = \phi,$$
$$b_x^{\text{TM}} = \frac{i\omega}{\nu^2 c^2}\frac{\partial \phi}{\partial y}, \quad b_y^{\text{TM}} = -\frac{i\omega}{\nu^2 c^2}\frac{\partial \phi}{\partial x}, \quad b_z^{\text{TM}} = 0. \tag{11.129}$$

The corresponding results for TE-waves are

$$\frac{\partial^2 \psi}{\partial x^2} + \frac{\partial^2 \psi}{\partial y^2} + \nu^2 \psi = 0, \tag{11.130}$$

$$e_x^{\text{TE}} = -\frac{i\omega}{\nu^2}\frac{\partial \psi}{\partial y}, \quad e_y^{\text{TE}} = \frac{i\omega}{\nu^2}\frac{\partial \psi}{\partial x}, \quad e_z^{\text{TE}} = 0,$$
$$b_x^{\text{TE}} = -\frac{i\gamma}{\nu^2}\frac{\partial \psi}{\partial x}, \quad b_y^{\text{TE}} = -\frac{i\gamma}{\nu^2}\frac{\partial \psi}{\partial y}, \quad b_z^{\text{TE}} = \psi. \tag{11.131}$$

We now have to apply the usual boundary conditions

$$\hat{\boldsymbol{n}}\cdot\boldsymbol{B} = 0, \quad \hat{\boldsymbol{n}}\times\boldsymbol{E} = \boldsymbol{0},$$

where $\hat{\boldsymbol{n}}$ stands for the unit normal to the bounding surface drawn into the field, see Fig. 11.10, which must apply to any field at the surface of a perfect conductor.

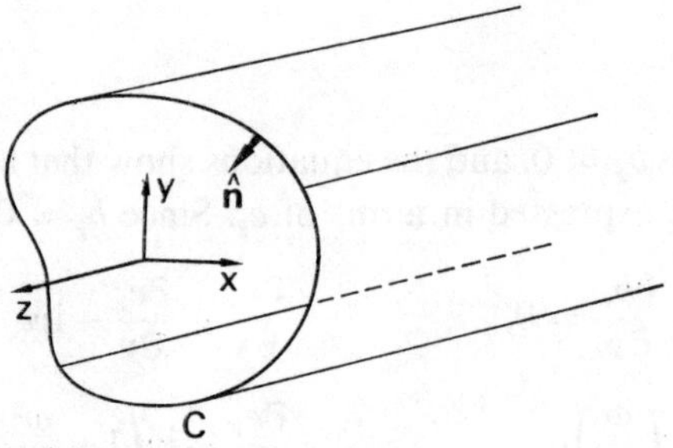

FIG. 11.10

First we consider TM-waves. On the boundary, with the unit normal given by $\hat{\boldsymbol{n}} = \boldsymbol{i}l+\boldsymbol{j}m$ $(l^2+m^2 = 1)$,

$$\hat{\boldsymbol{n}}\cdot\boldsymbol{B} = \frac{i\omega}{\nu^2c^2}(\boldsymbol{i}l+\boldsymbol{j}m)\cdot\left(-\boldsymbol{i}\frac{\partial\phi}{\partial y}+\boldsymbol{j}\frac{\partial\phi}{\partial x}\right) = 0.$$

Therefore
$$-l\frac{\partial\phi}{\partial y}+m\frac{\partial\phi}{\partial x} = 0. \tag{11.132}$$

Also

$$\hat{\boldsymbol{n}}\times\boldsymbol{E} = (\boldsymbol{i}l+\boldsymbol{j}m)\times\left\{-\frac{i\gamma}{\nu^2}\left(\boldsymbol{i}\frac{\partial\phi}{\partial x}+\boldsymbol{j}\frac{\partial\phi}{\partial y}\right)+\boldsymbol{k}\phi\right\}$$

$$= \boldsymbol{i}m\phi-\boldsymbol{j}l\phi-\frac{i\gamma}{\nu^2}\boldsymbol{k}\left(l\frac{\partial\phi}{\partial y}-m\frac{\partial\phi}{\partial x}\right) = \boldsymbol{0}. \tag{11.133}$$

Both these conditions are satisfied if $\phi = 0$ everywhere on the boundary curve C in the xy-plane. Since ϕ depends only on x and y,

$$\textbf{grad}\ \phi = \boldsymbol{i}\frac{\partial\phi}{\partial x}+\boldsymbol{j}\frac{\partial\phi}{\partial y}.$$

Therefore

$$\hat{\boldsymbol{n}}\times\textbf{grad}\ \phi = (\boldsymbol{i}l+\boldsymbol{j}m)\times\left(\boldsymbol{i}\frac{\partial\phi}{\partial x}+\boldsymbol{j}\frac{\partial\phi}{\partial y}\right) = \boldsymbol{k}\left(l\frac{\partial\phi}{\partial y}-m\frac{\partial\phi}{\partial x}\right).$$

But since $\phi = 0$ on C, **grad** ϕ is parallel to the normal $\hat{\boldsymbol{n}}$, and their vector product is zero. Hence, to determine the TM-waves given by eqns. (11.128–129) we must solve (11.128) subject to the condition $\phi = 0$ everywhere on the boundary C. This is an eigenvalue problem which has solutions ϕ_i only for certain eigenvalues ν_i; also these eigenfunctions ϕ_i satisfy an orthogonality condition

$$\iint \phi_i\phi_j\ \mathrm{d}x\ \mathrm{d}y = \delta_{ij}$$

where the integral is taken over the area enclosed by C. Further, an arbitrary solution satisfying the boundary conditions can always be written as a Fourier-type expansion $\phi = \sum_i a_i\phi_i$. By means of such an expansion we can obtain a TM-wave corresponding to any given field values at the end $z = 0$ of the tube.

From the derivation of (11,128) we have

$$\gamma^2 = \omega^2/c^2-\nu^2. \tag{11.134}$$

Because ν can only take one of a set of discrete values ν_i only certain values of γ are possible for a given frequency. Also, if attenuation of the wave is not to take place, γ must be real and therefore ω must exceed the value $c\nu_i$. Hence frequencies below $c\nu_i/(2\pi)$ cannot be transmitted down the tube in the mode corresponding to ϕ_i; there is a *cut-off frequency*. Note that different frequencies corresponding to a given mode ν_i travel with different velocities, and different modes with a given frequency also travel with different velocities. Hence, in addition to the "cut-off" effect, the "profile" of the signal is altered as it travels down the tube and distortion takes place. (This effect is familiar in acoustics where the sound of a voice after it has travelled along a pipe is very different from the original.)

The situation with TE-waves is closely similar. When we apply the boundary conditions to eqns. (11.131) we obtain

$$\hat{\boldsymbol{n}}\cdot\boldsymbol{B} = (\boldsymbol{i}l+\boldsymbol{j}m)\cdot\left\{-\frac{\mathrm{i}\gamma}{\nu^2}\left(\boldsymbol{i}\frac{\partial\psi}{\partial x}+\boldsymbol{j}\frac{\partial\psi}{\partial y}\right)+\boldsymbol{k}\psi\right\} = -\frac{\mathrm{i}\gamma}{\nu^2}\left(l\frac{\partial\psi}{\partial x}+m\frac{\partial\psi}{\partial y}\right) = 0,$$

$$\hat{\boldsymbol{n}}\times\boldsymbol{E} = (\boldsymbol{i}l+\boldsymbol{j}m)\times\left\{\frac{\mathrm{i}\omega}{\nu^2}\left(-\boldsymbol{i}\frac{\partial\psi}{\partial y}+\boldsymbol{j}\frac{\partial\psi}{\partial x}\right)\right\} = \frac{\mathrm{i}\omega}{\nu^2}\boldsymbol{k}\left(l\frac{\partial\psi}{\partial x}+m\frac{\partial\psi}{\partial y}\right) = \boldsymbol{0}.$$

These conditions are both satisfied if

$$l\frac{\partial\psi}{\partial x}+m\frac{\partial\psi}{\partial y} = \frac{\partial\psi}{\partial n} = 0, \tag{11.135}$$

where $\partial\psi/\partial n$ is the derivative of ψ along the normal to C. Therefore the possible TE-waves are obtained from the solution of eqn. (11.130) subject to the boundary condition $\partial\psi/\partial n = 0$ everywhere on C. This again is an eigenvalue problem which has a solution ψ_i only for certain, discrete values of $\nu = \nu_i$. The other considerations apply as for TM-waves.

The solution for TEM-waves is trivial unless the region in the xy-plane enclosed by C is multiply connected. The discussion of this case is given in the example below (p. 480).

The determination of the TM- or TE-waves depends first of all on the solution of eqns. (11.128) or (11.130) subject to the boundary conditions, followed by derivation of the fields from (11.129) or (11.131).

The mean energy-flow in the tube associated with either TM- or TE-waves is easily obtained as the real part of the complex Poynting vector

$$\boldsymbol{P} = \tfrac{1}{2}(\boldsymbol{E}\times\boldsymbol{H}^*) = (\boldsymbol{E}\times\boldsymbol{B}^*)/(2\mu_0).$$

Provided that ϕ, or ψ, obtained from the solution of (11.128) or (11.130), is real, the Poynting vector for TM-waves is

$$\boldsymbol{P} = \frac{1}{2\mu_0}\left\{-\frac{\mathrm{i}\omega}{\nu^2c^2}\phi\frac{\partial\phi}{\partial x}\boldsymbol{i}-\frac{\mathrm{i}\omega}{\nu^2c^2}\phi\frac{\partial\phi}{\partial y}\boldsymbol{j}+\frac{\omega\gamma}{\nu^4c^2}\left[\left(\frac{\partial\phi}{\partial x}\right)^2+\left(\frac{\partial\phi}{\partial y}\right)^2\right]\boldsymbol{k}\right\}.$$

So long as γ is real this shows a mean energy-flow only in the z-direction of amount

$$\frac{\omega\gamma}{2\nu^4c^4\mu_0}\left\{\left(\frac{\partial\phi}{\partial x}\right)^2+\left(\frac{\partial\phi}{\partial y}\right)^2\right\}.$$

If γ is imaginary, the mean energy flow is zero since $\boldsymbol{P}$ has no real part, and corresponds to the fact that, when $\gamma^2 < 0$, attenuation of the wave takes place. Similarly for TE-waves the mean energy flow is

$$\frac{\gamma\omega}{2\mu_0\nu^4}\left\{\left(\frac{\partial\psi}{\partial x}\right)^2+\left(\frac{\partial\psi}{\partial y}\right)^2\right\}$$

in the z-direction.

The energy in the field is distributed with mean density

$$\text{Re}\ \tfrac{1}{2}\{\tfrac{1}{2}\varepsilon_0\boldsymbol{E}\cdot\boldsymbol{E}^*+\tfrac{1}{2}\mu_0^{-1}\boldsymbol{B}\cdot\boldsymbol{B}^*\}.$$

In the TM-mode, the energy in unit length of the tube is therefore the real part of

$$W=\frac{1}{4}\iint\left\{\varepsilon_0\left[\frac{\gamma^2}{\nu^4}\left(\frac{\partial\phi}{\partial x}\right)^2+\frac{\gamma^2}{\nu^4}\left(\frac{\partial\phi}{\partial y}\right)^2+\phi^2\right]\right.$$
$$\left.+\mu_0^{-1}\frac{\omega^2}{\nu^4c^4}\left[\left(\frac{\partial\phi}{\partial x}\right)^2+\left(\frac{\partial\phi}{\partial y}\right)^2\right]\right\}\mathrm{d}x\,\mathrm{d}y,$$

where the integral is taken over the area of cross-section of the tube. This is

$$W=\frac{\varepsilon_0}{4\nu^4}\iint\left\{\left(\gamma^2+\frac{\omega^2}{c^2}\right)\left[\left(\frac{\partial\phi}{\partial x}\right)^2+\left(\frac{\partial\phi}{\partial y}\right)^2\right]+\nu^4\phi^2\right\}\mathrm{d}x\,\mathrm{d}y.$$

From Green's theorem and eqn. (11.128) we show that, since $\phi = 0$ on C,

$$0=\oint_C\phi\frac{\partial\phi}{\partial n}\,\mathrm{d}s=\iint\{\phi\,\nabla^2\phi+(\nabla\phi)^2\}\,\mathrm{d}x\,\mathrm{d}y=\iint\{-\nu^2\phi^2+(\nabla\phi)^2\}\,\mathrm{d}x\,\mathrm{d}y.$$

Therefore

$$\iint\left\{\left(\frac{\partial\phi}{\partial x}\right)^2+\left(\frac{\partial\phi}{\partial y}\right)^2\right\}\mathrm{d}x\,\mathrm{d}y=\nu^2\iint\phi^2\,\mathrm{d}x\,\mathrm{d}y$$

so that

$$W=\frac{\varepsilon_0}{4\nu^4}\iint\left(\gamma^2+\frac{\omega^2}{c^2}+\nu^2\right)\left[\left(\frac{\partial\phi}{\partial x}\right)^2+\left(\frac{\partial\phi}{\partial y}\right)^2\right]\mathrm{d}x\,\mathrm{d}y$$
$$=\frac{\varepsilon_0\omega^2}{2\nu^4c^2}\iint\left[\left(\frac{\partial\phi}{\partial x}\right)^2+\left(\frac{\partial\phi}{\partial y}\right)^2\right]\mathrm{d}x\,\mathrm{d}y.$$

The rate of flow of energy down the tube is

$$P = \frac{\omega\gamma}{2\nu^4c^2\mu_0}\iint\left[\left(\frac{\partial\phi}{\partial x}\right)^2+\left(\frac{\partial\phi}{\partial y}\right)^2\right]dx\,dy = Wv_g,$$

where v_g, the "group-velocity" is the velocity with which energy is apparently propagated along the tube and differs from the phase-velocity v_p. Note that

$$v_g = \frac{\omega\gamma}{2\nu^4c^2\mu_0}\,\frac{2\nu^4c^2}{\varepsilon_0\omega^2} = \frac{\gamma}{\varepsilon_0\mu_0\omega} = \frac{c^2}{v_p}, \quad \text{i.e.} \quad v_pv_g = c^2.$$

(The phenomenon of "group-velocity" arises whenever the phase-velocity of a wave depends on the frequency. This dependence on frequency occurs here.)

Example. The discussion for TEM-waves.

These waves correspond to case 3 on p. 475 with

$$(\boldsymbol{e}\cdot\boldsymbol{k}) = 0 = e_z, \qquad (\boldsymbol{b}\cdot\boldsymbol{k}) = 0 = b_z.$$

The remaining equations from (11.123–126c) now give

$$\frac{\partial e_x}{\partial x}+\frac{\partial e_y}{\partial y} = 0, \quad (1) \qquad \frac{\partial b_x}{\partial x}+\frac{\partial b_y}{\partial y} = 0, \quad (2)$$

$$\frac{\partial e_x}{\partial y}-\frac{\partial e_y}{\partial x} = 0, \quad (3) \qquad \frac{\partial b_x}{\partial y}-\frac{\partial b_y}{\partial x} = 0, \quad (4)$$

$$e_y = -(\omega/\gamma)b_x, \quad (5) \qquad b_y = (\omega/\gamma c^2)e_x, \quad (6)$$

$$e_x = (\omega/\gamma)b_y, \quad (7) \qquad b_x = -(\omega/\gamma c^2)e_y. \quad (8)$$

Clearly, eqns. (5)–(8) are satisfied, apart from the trivial case $e_x = e_y = b_x = b_y = 0$, only if

$$\frac{\omega}{\gamma} = \frac{\gamma c}{\omega}, \quad \text{i.e.} \quad \gamma^2 = \frac{\omega^2}{c^2}. \tag{9}$$

This is equivalent to $\nu^2=0$ in (11.134). This also implies that eqns. (1) and (3) are the same as eqns. (4) and (2) respectively. We can satisfy (4) by writing

$$b_x = \frac{\partial f}{\partial x}, \quad b_y = \frac{\partial f}{\partial y} \quad (\text{i.e. } \boldsymbol{b} = \mathbf{grad}\, f)$$

where

$$\frac{\partial^2 f}{\partial x^2}+\frac{\partial^2 f}{\partial y^2} = 0. \tag{10}$$

Here f is a single-valued function only if the region enclosed by C in the xy-plane is singly connected. Otherwise f is a cyclic function.

Now the imposition of the boundary conditions on the solution of (10), viz. $f = 0$, or $\partial f/\partial x = 0$ on C, implies that the only solution, *in a singly connected space*, is $f = 0$. Hence there can be no waves of type TEM in a single hollow pipe. Such waves can be propagated in a coaxial cable, where one conductor encloses the other. In this case f is a cyclic function and non-zero fields exist. We cannot pursue the discussion of propagation along coaxial cables here; for this the reader should consult special treatises. If the field is established in the space *outside* a conducting tube, the area is doubly connected and the TEM mode can exist (see § 11.8).

Rectangular waveguides

We develop the solution for the special case of a tube with rectangular cross-section shown in Fig. 11.11. We will consider only TM-waves.

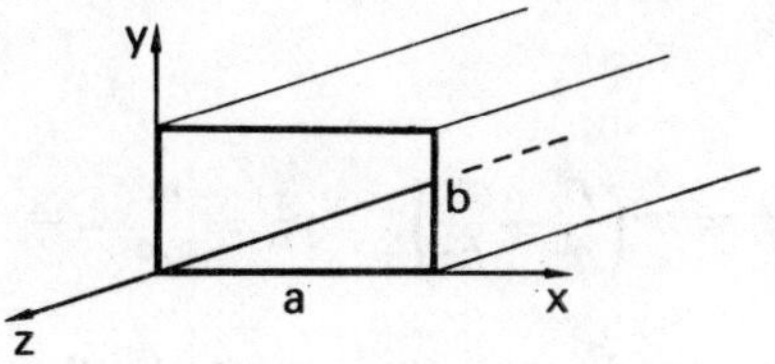

FIG. 11.11

We have to solve (11.128), viz.

$$\frac{\partial^2 \phi}{\partial x^2}+\frac{\partial^2 \phi}{\partial y^2}+\nu^2\phi = 0,$$

subject to the boundary condition $\phi = 0$ on all boundaries $x = 0$, $x = a$, $y = 0$, $y = b$. The solution is found by the method of separation of variables which gives

$$\phi_{rs} = A_{rs} \sin\left(\frac{r\pi x}{a}\right) \sin\left(\frac{s\pi y}{b}\right),$$

where the eigenvalues are given by

$$\nu^2 = \nu_{rs}^2 = \frac{r^2\pi^2}{a^2}+\frac{s^2\pi^2}{b^2},$$

and r, s $(= 1, 2, \ldots)$ must take integral values in order to make $\phi_{rs} = 0$ on all boundaries. (It is here obviously convenient to use two suffixes r, s, instead of the single suffix i, when we label the various eigenvalues and eigenfunctions.)

The various components of the field are given by

$$E_x^{TM} = -\frac{i\gamma_{rs}}{\nu_{rs}^2} A_{rs}\left(\frac{r\pi}{a}\right) \cos\left(\frac{r\pi x}{a}\right) \sin\left(\frac{s\pi y}{b}\right) \exp\{i(\omega t-\gamma_{rs}z)\},$$

$$E_y^{TM} = -\frac{i\gamma_{rs}}{\nu_{rs}^2} A_{rs}\left(\frac{s\pi}{b}\right) \sin\left(\frac{r\pi x}{a}\right) \cos\left(\frac{s\pi y}{b}\right) \exp\{i(\omega t-\gamma_{rs}z)\},$$

$$E_z^{TM} = A_{rs} \sin\left(\frac{r\pi x}{a}\right) \sin\left(\frac{s\pi y}{b}\right) \exp\{i(\omega t - \gamma_{rs} z)\}, \tag{11.136}$$

$$B_x^{TM} = \frac{i\omega}{\nu_{rs}^2 c^2} A_{rs}\left(\frac{s\pi}{b}\right) \sin\left(\frac{r\pi x}{a}\right) \cos\left(\frac{s\pi y}{b}\right) \exp\{i(\omega t - \gamma_{rs} z)\},$$

$$B_y^{TM} = -\frac{i\omega}{\nu_{rs}^2 c^2} A_{rs}\left(\frac{r\pi}{a}\right) \cos\left(\frac{r\pi x}{a}\right) \sin\left(\frac{s\pi y}{b}\right) \exp\{(i\omega t - \gamma_{rs} z)\},$$

$$B_z^{TM} = 0,$$

where

$$\nu_{rs}^2 = \pi^2\left(\frac{r^2}{a^2} + \frac{s^2}{b^2}\right), \quad \gamma_{rs}^2 = \frac{\omega^2}{c^2} - \nu_{rs}^2.$$

These results, as they stand, are not particularly illuminating but we can obtain certain other results from them.

The simplest mode of TM-wave occurs with $r = s = 1$ and

$$\nu_{11}^2 = \pi^2\left(\frac{1}{a^2} + \frac{1}{b^2}\right)$$

and the "cut-off" frequency is

$$\frac{\omega_0}{2\pi} = \frac{c\nu_{11}}{2\pi} = \frac{c}{2}\left(\frac{1}{a^2} + \frac{1}{b^2}\right)^{1/2}.$$

A wave of this frequency has a wavelength *in free space* of

$$\lambda_0 = \frac{2\pi c}{\omega_0} = 2\left(\frac{1}{a^2} + \frac{1}{b^2}\right)^{-1/2}.$$

This indicates that the wavelength *in free space* of radiation which can pass down the tube without attenuation must have the same order of magnitude, or less, as the transverse dimensions of the tube. For a square section of side 5 cm the maximum wavelength transmitted is 7·1 cm. (We give the wavelength "in free space" because inside the tube the wavelength is given by $2\pi/\gamma$; but for the critical frequency of cut-off $\gamma = 0$ and the wavelength in the tube is infinite.)

The corresponding solutions for TE-waves come from the solution of the equation

$$\frac{\partial^2 \psi}{\partial x^2} + \frac{\partial^2 \psi}{\partial y^2} + \nu^2 \psi = 0$$

subject to the conditions:

$$\frac{\partial \psi}{\partial x} = 0 \quad \text{on} \quad x = 0, \quad a; \qquad \frac{\partial \psi}{\partial y} = 0 \quad \text{on} \quad y = 0, \quad b.$$

The solutions, obtained by the method of separation of variables, are

$$\psi_{rs} = B_{rs} \cos\left(\frac{r\pi x}{a}\right) \cos\left(\frac{s\pi y}{b}\right),$$

$$\nu_{rs}^2 = \pi^2\left(\frac{r^2}{a^2} + \frac{s^2}{b^2}\right), \quad \gamma_{rs}^2 = \frac{\omega^2}{c^2} - \nu_{rs}^2.$$

The simplest modes correspond here to

$$\psi_{10} = B_{10} \cos\left(\frac{\pi x}{a}\right), \quad \psi_{01} = B_{01} \cos\left(\frac{\pi y}{b}\right).$$

The critical wavelengths of cut-off are $2a$, $2b$ having frequencies $c/(2a)$, $c/(2b)$ respectively.

Example 1. The parallel plate waveguide

This is a waveguide with walls consisting of the planes $x = 0$, $x = a$ only in which propagation takes place in the z-direction, but there is no dependence on y. We include this because the solution enables us to understand how the more complicated phenomena with rectangular, and other shaped, waveguides arise physically.

Because there is no dependence on y we start, for TM-waves, with the solution of

$$\frac{\partial^2 \phi}{\partial x^2} + \nu^2 \phi = 0; \qquad \phi = 0 \quad \text{on} \quad x = 0, \ a.$$

This is given by

$$\phi_r = A_r \sin\left(\frac{r\pi x}{a}\right), \qquad r = 1, 2, \ldots,$$

$$\nu_r = \frac{r\pi}{a}, \quad \gamma_r^2 = \frac{\omega^2}{c^2} - \frac{r^2\pi^2}{a^2}. \tag{1}$$

The field components are, from eqn. (11.129),

$$E_x = -\frac{i\gamma_r}{\nu_r} A_r \cos\left(\frac{r\pi x}{a}\right) \exp\{i(\omega t - \gamma_r z)\}, \quad E_y = 0,$$

$$E_z = A_r \sin\left(\frac{r\pi x}{a}\right) \exp\{i(\omega t - \gamma_r z)\}, \tag{2}$$

$$B_x = 0, \quad B_y = \frac{i\omega}{\nu_r c} A_r \cos\left(\frac{r\pi x}{a}\right) \exp\{i(\omega t - \gamma_r z)\}, \quad B_z = 0.$$

From (1) we may put

$$\nu_r = \frac{r\pi}{a} = \frac{\omega}{c}\sin\alpha, \qquad \gamma_r = \frac{\omega}{c}\cos\alpha,$$

and insert
$$\cos\left(\frac{r\pi x}{a}\right) = \frac{1}{2}\exp\left(\frac{ir\pi x}{a}\right) + \frac{1}{2}\exp\left(-\frac{ir\pi x}{a}\right)$$
$$= \frac{1}{2}\left\{\exp\left(\frac{ix\omega\sin\alpha}{a}\right) + \exp\left(-\frac{ix\omega\sin\alpha}{a}\right)\right\}$$

into the field components. We obtain, for example,

$$E_x = -(\mathrm{i}/2)(\cot\alpha)A_r[\exp\{(\mathrm{i}\omega/c)(ct - x\sin\alpha - z\cos\alpha)\} + \exp\{(-\mathrm{i}\omega/c)(ct + x\sin\alpha - z\cos\alpha)\}]$$

with similar expressions for E_z, B_y. These expressions represent the combination of two harmonic waves with constant amplitudes propagated with phase-velocity c in the directions $-\boldsymbol{i}\sin\alpha + \boldsymbol{k}\cos\alpha$, $\boldsymbol{i}\sin\alpha + \boldsymbol{k}\cos\alpha$ (see Fig. 11.12).

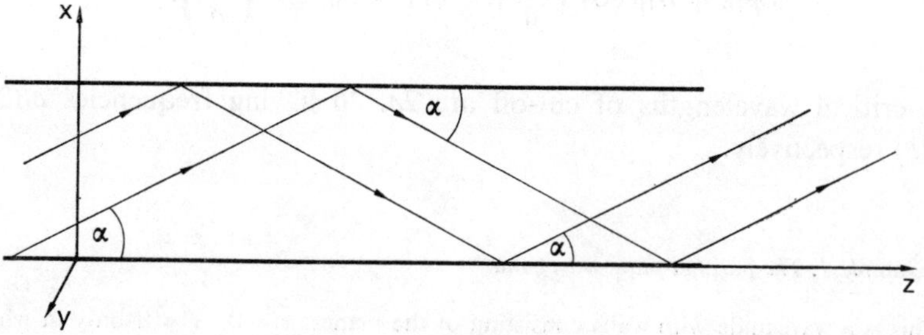

FIG. 11.12

The expressions (2) correspond to a wave (with variable amplitude) propagated in the z-direction, having frequency $\omega/(2\pi)$, wavelength $2\pi/\gamma_r$, and phase velocity ω/γ_r. These are seen to be the resultant of plane waves, with constant amplitude, reflected obliquely to and fro between the plates, each wave having frequency $\omega/(2\pi)$, wavelength $2\pi c/\omega$ (the "free space value") and phase velocity c. The "cut-off" wavelength corresponds to an angle $\alpha = \pi/2$ in which the waves are reflected directly across perpendicular to the plates and so transmit no energy in the z-direction. In this solution (a TM-wave) $\boldsymbol{B}$ is always transverse, while $\boldsymbol{E}$ has a component in the z-direction; the corresponding TE-wave has $E_x = E_z = 0$ and $\boldsymbol{B}$ has a z-component, the solution being given by reflection as before.

We can now understand how the reflection from another pair of faces in the rectangular waveguide, or from the curved surface of some other shaped tube, builds up the more complicated modes in the general investigation.

Example 2. Show that the formulae

$$\boldsymbol{E} = \textbf{curl curl } \boldsymbol{S}, \qquad \boldsymbol{B} = \frac{1}{c^2}\textbf{ curl}\left(\frac{\partial \boldsymbol{S}}{\partial t}\right)$$

for the electric and magnetic vectors satisfy Maxwell's equations in empty space provided that in a rectangular coordinate system the vector function $\boldsymbol{S}$ is a solution of the wave

equation
$$c^2\,\nabla^2\boldsymbol{S} = \partial^2\boldsymbol{S}/\partial t^2.$$

Verify that $A\boldsymbol{k}\sin\alpha x\sin(\omega t-\gamma z)$ is a possible form of $\boldsymbol{S}$, where A, α, γ are real constants, x, y, z rectangular coordinates and $\boldsymbol{k}$ is a unit vector along the z-axis, if α, γ, and ω satisfy a certain condition.

Find the condition that such a solution should represent a field between two perfectly conducting planes $x = 0$ and $x = a$, and prove that if the field is that of a wave propagated along the z-axis then $\omega > n\pi c/a$, where n is an integer.

Show also that the time average of the energy-flow is parallel to the z-axis and of amount $\omega\gamma a\alpha^2 A^2/(4\mu_0)$ per unit width.

With the suggested forms for the field vectors the equations
$$\operatorname{div}\boldsymbol{E} = 0,\qquad \operatorname{div}\boldsymbol{B} = 0,\qquad \frac{1}{\mu_0}\,\mathbf{curl}\,\boldsymbol{B}-\varepsilon_0\frac{\partial\boldsymbol{E}}{\partial t} = \boldsymbol{0},$$
are satisfied identically. The remaining equation gives
$$\mathbf{curl}\,\boldsymbol{E}+\frac{\partial\boldsymbol{B}}{\partial t} = \mathbf{curl}\left\{\mathbf{curl}\;\mathbf{curl}\,\boldsymbol{S}+\frac{1}{c^2}\frac{\partial^2\boldsymbol{S}}{\partial t^2}\right\}$$
$$= \mathbf{curl}\left\{\mathbf{grad}\operatorname{div}\boldsymbol{S}-\nabla^2\boldsymbol{S}+\frac{1}{c^2}\frac{\partial^2\boldsymbol{S}}{\partial t^2}\right\}.$$

Since $\mathbf{curl}\;\mathbf{grad}\;u \equiv \boldsymbol{0}$, where in this case $u = \operatorname{div}\boldsymbol{S}$, the fourth of Maxwell's equations is satisfied when $\boldsymbol{S}$ satisfies the stated relation.

When
$$\boldsymbol{S} = \boldsymbol{k}A\sin(\alpha x)\sin(\omega t-\gamma z),$$
$$\nabla^2\boldsymbol{S} = \boldsymbol{k}(-\alpha^2-\gamma^2)A\sin(\alpha x)\sin(\omega t-\gamma z),$$
$$\frac{\partial^2\boldsymbol{S}}{\partial t^2} = -\omega^2\boldsymbol{k}A\sin(\alpha x)\sin(\omega t-\gamma z).$$

Therefore
$$\alpha^2+\gamma^2 = \omega^2/c^2. \tag{1}$$

Also
$$\mathbf{curl}\,\boldsymbol{S} = -\boldsymbol{j}\alpha A\cos(\alpha x)\sin(\omega t-\gamma z),$$
$$\mathbf{curl}\;\mathbf{curl}\,\boldsymbol{S} = -\boldsymbol{i}\gamma\alpha A\cos(\alpha x)\cos(\omega t-\gamma z)+\boldsymbol{k}\alpha^2A\sin(\alpha x)\sin(\omega t-\gamma z) = \boldsymbol{E}.$$

Therefore
$$\boldsymbol{B} = -\boldsymbol{j}\alpha\omega A\cos(\alpha x)\cos(\omega t-\gamma z).$$

The boundary conditions to be satisfied on $x = 0$, $x = a$ are $\boldsymbol{i}\cdot\boldsymbol{B} = 0$, $\boldsymbol{i}\times\boldsymbol{E} = \boldsymbol{0}$. The first of these is identically satisfied, and the second is satisfied for $x = 0$. When $x = a$ we must have the condition $\sin(\alpha a) = 0$ so that
$$\alpha = n\pi/a,\qquad (n = 1, 2, \ldots).$$

For propagation of a wave in the z-direction γ must be real and $\gamma^2 > 0$, and so, from (1),
$$\frac{\omega}{c} > \frac{n\pi}{a}.$$

The instantaneous value of the energy-flow is given by the Poynting vector $\boldsymbol{P}$, where
$$\mu_0\boldsymbol{P} = \mu_0(\boldsymbol{E}\times\boldsymbol{H}) = (\boldsymbol{E}\times\boldsymbol{B})$$
$$= \boldsymbol{i}\omega\alpha^3A^2\sin(\alpha x)\cos(\alpha x)\sin(\omega t-\gamma z)\cos(\omega t-\gamma z)+\boldsymbol{k}\alpha^2\gamma\omega A^2\cos^2(\alpha x)\cos^2(\omega t-\gamma z).$$

Because of the factor $\sin(\omega t-\gamma z)\cos(\omega t-\gamma z)$ the time-average of the x-component of $\boldsymbol{P}$ is zero, and the z-component has the mean value $\frac{1}{2}[\gamma\omega\alpha^2A^2/\mu_0]\cos^2(\alpha x)$. Therefore the total flow in the z-direction across unit width in the y-direction between the plates is

$$\frac{\gamma\omega A^2\alpha^2}{2\mu_0}\int_0^a \cos^2(\alpha x)\,dx = \frac{\gamma\omega\alpha^2A^2a}{4\mu_0}.$$

Example 3. Electromagnetic waves travel along a perfectly conducting guide of rectangular cross-section bounded by $x = 0$, $x = a$; $y = 0$, $y = b$. Prove that, for the E_{mn} mode, the component of the electric field parallel to the axis of the guide can be expressed in the form

$$E_z = \sin(m\pi x/a)\sin(n\pi y/b)\exp\{i(\omega t-\beta z)\},$$

and determine the other components of $\boldsymbol{E}$ and $\boldsymbol{B}$.

Determine the critical wavelength and the wavelength in the guide corresponding to a wavelength λ in the unbounded medium.

Find the limits between which the length of the side of a guide of square cross-section must lie in order that only the E_{11} mode is transmitted without attenuation.

This case was discussed in the text, with a slight difference of notation, viz. r, s were used instead of m, n here, and γ_{rs} was used instead of β here. With these changes the field vector components are given in eqns. (11.136).

The critical wavelength for any mode of oscillation is the wavelength λ_c in free space which a wave having the "cut-off" frequency must have. The "cut-off" frequency for the mode given here is $\omega/(2\pi)$ where ω/c has the value making γ_{rs} (or β) vanish, i.e.

$$\lambda_c = \frac{2\pi c}{\omega} = 2\left(\frac{m^2}{a^2}+\frac{n^2}{b^2}\right)^{-1/2}.$$

For other frequencies the wavelength in the tube is λ_t where $\lambda_t = 2\pi/\beta$, and the free space wavelength is $\lambda = 2\pi c/\omega$. Hence, from the relation

$$\beta^2 = \frac{\omega^2}{c^2}-\pi^2\left(\frac{m^2}{a^2}+\frac{n^2}{b^2}\right)$$

we deduce that

$$\frac{1}{\lambda_t^2} = \frac{1}{\lambda^2}-\frac{1}{4}\left(\frac{m^2}{a^2}+\frac{n^2}{b^2}\right).$$

For a square waveguide and an oscillation in the E_{11} mode, $a = b$, $m = n = 1$. Hence the frequency of the oscillation must exceed the cut-off frequency. Therefore

$$\frac{\omega^2}{c^2} > \frac{2\pi^2}{a^2}, \quad \text{i.e.} \quad a > \sqrt{(2)}\pi c/\omega.$$

The modes with the nearest cut-off frequency to this value are the E_{21}, E_{12} modes with $\nu_{12}^2 = \nu_{21}^2 = 5\pi^2/a^2$. Hence, if these modes are not to be transmitted, the frequency must lie below the "cut-off" value for these modes, i.e.

$$\frac{\omega^2}{c^2} < \frac{5\pi^2}{a^2}, \quad \text{i.e.} \quad a < \sqrt{(5)}\pi c/\omega.$$

We thus obtain the limits on a to be

$$(\pi c/\omega)\sqrt{2} < a < (\pi c/\omega)\sqrt{5},$$

where $\omega/(2\pi)$ is the frequency of the wave being used.

Example 4. A long circular cylinder, of uniform material, of radius a and permittivity ε and permeability μ, is embedded in a perfect conductor, the axis of the cylinder being the z-axis. If the magnetic field has no component along this axis, and all other field quantities are of the form $R(r)\,\Theta(\theta)\exp\{i(\omega t-\beta z)\}$, where ω, β are constants and (r, θ, z) are cylindrical polar coordinates, show that $J_n(\nu a)=0$, where $J_n(x)$ is the Bessel function of the first kind, n being an integer, and

$$\nu^2 = \left(\frac{\omega}{v}\right)^2-\beta^2,$$

where v is the phase velocity of electromagnetic waves in the medium.

Show that, for a given n, there is a critical value ω_0 which ω must exceed if waves of this type are to be transmitted in the cylinder.

We include this example as an indication of how non-rectangular wave guides may be discussed, and to show how there is a corresponding pattern of possible modes of oscillation.

We return to eqns. (11.123), (11.125) and (11.126) where the field vectors are expressed in terms of $\boldsymbol{e}$, $\boldsymbol{b}$, which depend only on the position of the field point in the cross-section of the waveguide, the dependence on z and t being included in the factor $\exp\{i(\omega t-\beta z)\}$. [Here β replaces γ of the previous discussion.] Since we use cylindrical polar coordinates (r, θ, z) (suggested by the form of the boundary) we now regard $\boldsymbol{e}$, $\boldsymbol{b}$ as functions only of r, θ.

Since we are given that the magnetic field has no component along the z-axis, we put $\boldsymbol{b}\cdot\boldsymbol{k} = b_z = 0$ and take $\boldsymbol{e}\cdot\boldsymbol{k} = \phi$. We find, as before, that all field quantities can be expressed in terms of ϕ. The form taken by Maxwell's equations now is [we replace c^2 by $1/(\mu\varepsilon)$ since the medium is not a vacuum]:

$$\operatorname{div}\boldsymbol{e}-i\beta(\boldsymbol{k}\cdot\boldsymbol{e}) = 0; \qquad \frac{1}{r}\left\{\frac{\partial}{\partial r}(re_r)+\frac{\partial}{\partial\theta}(e_\theta)\right\} = i\beta\phi$$

$$\operatorname{div}\boldsymbol{b}-i\beta(\boldsymbol{k}\cdot\boldsymbol{b}) = 0; \qquad \frac{1}{r}\left\{\frac{\partial}{\partial r}(rb_r)+\frac{\partial}{\partial\theta}(b_\theta)\right\} = 0.$$

curl $\boldsymbol{e}-i\beta(\boldsymbol{k}\times\boldsymbol{e})+i\omega\boldsymbol{b} = \boldsymbol{0}$:

$$\frac{1}{r}\left\{\frac{\partial}{\partial\theta}(\phi)-\frac{\partial}{\partial z}(re_\theta)\right\} = i\beta(-e_\theta)-i\omega b_r; \qquad \frac{1}{r}\frac{\partial\phi}{\partial\theta} = -i\beta e_\theta-i\omega b_r;$$

$$\frac{\partial}{\partial z}(e_r)-\frac{\partial}{\partial r}(\phi) = i\beta(e_r)-i\omega b_\theta; \qquad \frac{\partial\phi}{\partial r} = -i\beta e_r+i\omega b_\theta;$$

$$\frac{1}{r}\left\{\frac{\partial}{\partial r}(re_\theta)-\frac{\partial}{\partial\theta}(e_r)\right\} = 0; \qquad \frac{\partial e_\theta}{\partial r}+\frac{e_\theta}{r}-\frac{1}{r}\frac{\partial e_r}{\partial\theta} = 0;$$

curl $\boldsymbol{b}-i\beta(\boldsymbol{k}\times\boldsymbol{b})-i\omega\mu\varepsilon\boldsymbol{e} = \boldsymbol{0}$:

$$\frac{1}{r}\left\{\frac{\partial}{\partial\theta}(0)-\frac{\partial}{\partial z}(rb_\theta)\right\} = i\beta(-b_\theta)+i\omega\mu\varepsilon e_r; \qquad b_\theta = (\omega\mu\varepsilon/\beta)e_r;$$

$$\frac{\partial}{\partial z}(b_r)-\frac{\partial}{\partial r}(0) = i\beta(b_r)+i\omega\mu\varepsilon e_\theta; \qquad b_r = -(\omega\mu\varepsilon/\beta)e_\theta;$$

$$\frac{1}{r}\left\{\frac{\partial}{\partial r}(rb_\theta)-\frac{\partial}{\partial\theta}(b_r)\right\} = i\omega\mu\varepsilon\phi; \qquad \frac{\partial b_\theta}{\partial r}+\frac{b_\theta}{r}-\frac{1}{r}\frac{\partial b_r}{\partial\theta} = i\omega\mu\varepsilon\phi.$$

After some manipulations these equations lead to the following results, which correspond to eqns. (11.128–129),

$$\frac{\partial^2\phi}{\partial r^2}+\frac{1}{r}\frac{\partial\phi}{\partial r}+\frac{1}{r^2}\frac{\partial^2\phi}{\partial\theta^2}=(\beta^2-\omega^2\mu\varepsilon)\phi=-\nu^2\phi, \tag{1}$$

$$e_r=-\frac{i\beta}{\nu^2}\frac{\partial\phi}{\partial r},\qquad e_\theta=-\frac{i\beta}{\nu^2}\frac{1}{r}\frac{\partial\phi}{\partial\theta},\qquad e_z=\phi, \tag{2}$$

$$b_r=\frac{i\omega\mu\varepsilon}{\nu^2}\frac{1}{r}\frac{\partial\phi}{\partial\theta},\qquad b_\theta=-\frac{i\omega\mu\varepsilon}{\nu^2}\frac{\partial\phi}{\partial r},\qquad b_z=0. \tag{3}$$

Since the boundary of the field is a perfect conductor, eqn. (1) must be subject to the boundary condition $\phi=0$.

When we seek a solution of (1) in the form given we obtain

$$\frac{1}{R}\left(R''+\frac{1}{r}R'\right)+\frac{\Theta''}{r^2\Theta}=-\nu^2.$$

If Θ is to be a single-valued function of position we must have

$$\frac{\Theta''}{\Theta}=-n^2,\qquad \Theta=A\cos(n\theta)+B\sin(n\theta),\qquad (n=0,1,2,\ldots),$$

so that

$$r^2R''+rR'+(\nu^2r^2-n^2)R=0.$$

This is Bessel's equation of order n with the general solution

$$R=C_1J_n(\nu r)+C_2Y_n(\nu r).$$

Because $Y_n(\nu r)\to-\infty$ as $r\to 0$ we must choose $C_2=0$ in order to make R (and therefore ϕ) finite everywhere inside the waveguide. It follows that

$$R(r)=C_1J_n(\nu r).$$

The boundary condition at the metal surface requires that $J_n(\nu a)=0$, and implies that ν must take one of a discrete set of values ν_i.

The velocity of propagation of the waves is given by ω/β where

$$\nu_i^2=\omega^2\mu\varepsilon-\beta^2=(\omega/v)^2-\beta^2.$$

As in the discussion in the text, if the wave is to be transmitted along the guide, β must be real and so ω must exceed a critical value given by $\omega_0=v\nu_i$.

Example 5. Verify that Maxwell's equations for free space are satisfied by

$$\boldsymbol{B}=\frac{1}{c^2}\frac{\partial}{\partial t}\,\mathbf{grad}\,\phi\times\boldsymbol{k}$$

$$\boldsymbol{E}=-\boldsymbol{k}\frac{1}{c^2}\frac{\partial^2\phi}{\partial t^2}+\frac{\partial}{\partial z}\,\mathbf{grad}\,\phi$$

where $\boldsymbol{k}$ is the unit vector along the z-axis, and ϕ satisfies

$$\nabla^2\phi=\frac{1}{c^2}\frac{\partial^2\phi}{\partial t^2}.$$

Show that ϕ can be of the form

$$A \sin(\alpha x) \sin(\beta y) \cos(\gamma z) \cos(\omega t),$$

provided that the angular frequency ω and the constants α, β, γ satisfy a certain relation. Find the cartesian components of $\boldsymbol{E}$ in this case, and deduce that such a field can exist in the region $0 \leqslant x \leqslant l, 0 \leqslant y \leqslant l, 0 \leqslant z \leqslant l$, where the boundaries are perfectly conducting, if α, β, γ have suitably chosen values. If the field does not vanish identically, show that the least allowed value of ω is $(\pi c/l)\sqrt{2}$.

This is not strictly a problem on waveguides but its solution has many features in common with waveguide problems and shows how problems concerning electromagnetic fields inside cavities with conducting walls can be treated.

The verification of Maxwell's equations is a matter of manipulation of the vector differential operations. Since $\boldsymbol{k}$ is a constant vector, we can write the expression for $\boldsymbol{B}$ in the form

$$\boldsymbol{B} = \frac{1}{c^2}\frac{\partial}{\partial t}\,\textbf{curl}\,(\boldsymbol{k}\phi).$$

Hence div $\boldsymbol{B} = 0$, identically. Also

$$\frac{\partial}{\partial z}\,\textbf{grad}\,\phi = \textbf{grad}\left(\frac{\partial \phi}{\partial z}\right) = \textbf{grad}\,(\boldsymbol{k}\cdot\textbf{grad}\,\phi).$$

Therefore

$$\textbf{curl}\,\boldsymbol{E} = -\frac{1}{c^2}\frac{\partial^2}{\partial t^2}(\textbf{curl}\,\boldsymbol{k}\phi) = -\frac{\partial \boldsymbol{B}}{\partial t}$$

since

$$\textbf{curl grad}\left(\frac{\partial \phi}{\partial z}\right) \equiv \boldsymbol{0}.$$

Since the term $-\dfrac{\boldsymbol{k}}{c^2}\dfrac{\partial^2 \phi}{\partial t^2}$ has only a z-component,

$$\begin{aligned}\text{div}\,\boldsymbol{E} &= -\frac{\partial}{\partial z}\left(\frac{1}{c^2}\frac{\partial^2 \phi}{\partial t^2}\right) + \frac{\partial}{\partial z}(\text{div}\,\textbf{grad}\,\phi)\\ &= \frac{\partial}{\partial z}\left\{-\frac{1}{c^2}\frac{\partial^2 \phi}{\partial t^2} + \nabla^2\phi\right\} = 0\end{aligned}$$

by virtue of the condition satisfied by ϕ. Finally

$$\begin{aligned}\textbf{curl}\,\boldsymbol{H} &= \mu_0^{-1}\,\textbf{curl}\,\boldsymbol{B} = \varepsilon_0\frac{\partial}{\partial t}\,\textbf{curl curl}\,(\boldsymbol{k}\phi)\\ &= \varepsilon_0\frac{\partial}{\partial t}\{\textbf{grad}\,\text{div}\,(\boldsymbol{k}\phi) - \nabla^2(\boldsymbol{k}\phi)\}\\ &= \varepsilon_0\frac{\partial}{\partial t}\left\{\textbf{grad}\left(\frac{\partial \phi}{\partial z}\right) - \boldsymbol{k}\nabla^2\phi\right\} = \varepsilon_0\frac{\partial}{\partial t}\left\{-\frac{\boldsymbol{k}}{c^2}\frac{\partial^2 \phi}{\partial t^2} + \frac{\partial}{\partial z}(\textbf{grad}\,\phi)\right\}\\ &= \varepsilon_0\frac{\partial \boldsymbol{E}}{\partial t} = \frac{\partial \boldsymbol{D}}{\partial t}.\end{aligned}$$

Hence, all of Maxwell's equations for free space are satisfied.

Substitution of the given expression for ϕ into the wave equation shows that

$$\alpha^2 + \beta^2 + \gamma^2 = \omega^2/c^2.$$

The field components corresponding to the given ϕ are

$$E_x = -A\alpha\gamma \cos(\alpha x)\sin(\beta y)\sin(\gamma z)\cos(\omega t), \qquad \left[= \frac{\partial}{\partial z}\left(\frac{\partial \phi}{\partial x}\right)\right];$$

$$E_y = -A\beta\gamma \sin(\alpha x)\cos(\beta y)\sin(\gamma z)\cos(\omega t), \qquad \left[= \frac{\partial}{\partial z}\left(\frac{\partial \phi}{\partial y}\right)\right];$$

$$E_z = -A(\alpha^2+\beta^2)\sin(\alpha x)\sin(\beta y)\cos(\gamma z)\cos(\omega t), \qquad \left[= -\left(\frac{\partial^2 \phi}{\partial x^2}+\frac{\partial^2 \phi}{\partial y^2}\right)\right].$$

The value for E_z is obtained by noting that, since ϕ satisfies the wave equation,

$$E_z = -\frac{1}{c^2}\frac{\partial^2\phi}{\partial t^2}+\frac{\partial^2\phi}{\partial z^2} = -\frac{\partial^2\phi}{\partial x^2}-\frac{\partial^2\phi}{\partial y^2}.$$

Also, since $$\textbf{grad}\,\phi\times\boldsymbol{k} = \boldsymbol{i}\frac{\partial\phi}{\partial y}-\boldsymbol{j}\frac{\partial\phi}{\partial x},$$

$$B_x = \frac{1}{c^2}\frac{\partial^2\phi}{\partial y\,\partial t} = -\frac{\beta\omega}{c^2}A\sin(\alpha x)\cos(\beta y)\cos(\gamma z)\sin(\omega t),$$

$$B_y = -\frac{1}{c^2}\frac{\partial^2\phi}{\partial x\,\partial t} = \frac{\alpha\omega}{c^2}A\cos(\alpha x)\sin(\beta y)\cos(\gamma z)\sin(\omega t), \qquad B_z = 0.$$

When we substitute, in succession, $x = 0$, $y = 0$, $z = 0$ into these expressions we find that

$$\text{on } x = 0:\quad E_y = E_z = B_x = 0;$$
$$\text{on } y = 0:\quad E_x = E_z = B_y = 0;$$
$$\text{on } z = 0:\quad E_x = E_y = 0, \quad \text{with} \quad B_z = 0 \text{ identically.}$$

These results conform to the boundary condition

$$\hat{\boldsymbol{n}}\cdot\boldsymbol{B} = 0, \qquad \hat{\boldsymbol{n}}\times\boldsymbol{E} = \boldsymbol{0}$$

on each face. When we substitute $x = l$, $y = l$, $z = l$ we find that

$$\text{on } x = l:\quad E_y = E_z = B_x = 0 \quad \text{if} \quad \sin(\alpha l) = 0, \quad \alpha = m\pi/l;$$
$$\text{on } y = l:\quad E_x = E_z = B_y = 0 \quad \text{if} \quad \sin(\beta l) = 0, \quad \beta = n\pi/l;$$
$$\text{on } z = l:\quad E_x = E_y = 0 \quad \text{if} \quad \sin(\gamma l) = 0, \quad \gamma = p\pi/l;$$

where $m, n = 1, 2, 3, \ldots$, and $p = 0, 1, 2, \ldots$.
The extra possible value $p = 0$, whereas $m, n \neq 0$, occurs because the dependence of ϕ on z is through $\cos(\gamma z)$. Equation (1) now shows that

$$\omega^2/c^2 = \pi^2(m^2+n^2+p^2)/l^2,$$

and so the minimum possible value for ω is given by $m = 1 = n$, $p = 0$, i.e.

$$\omega_{\max} = (c\pi/l)\sqrt{2}.$$

Exercises 11.7

1. If the field resolutes are referred to rectangular axes *Oxyz*, show that there exists a solution (possessing continuous second derivatives) of the form

$$E_y = B_x = B_z = 0; \quad E_x = \partial\psi/\partial z, \quad E_z = -\partial\psi/\partial x, \quad c^2B_y = -\partial\psi/\partial t,$$

where ψ is a function of x, z and t satisfying

$$\frac{\partial^2\psi}{\partial x^2}+\frac{\partial^2\psi}{\partial z^2}=\frac{1}{c^2}\frac{\partial^2\psi}{\partial t^2}.$$

Assuming a solution of the form $\psi_0(z)\exp\{\mathrm{i}(\varkappa x-\nu t)\}$, of given frequency $\nu/2\pi$, where $\nu>\varkappa c$, find the form of the function $\psi_0(z)$ valid in the region $0\leqslant z\leqslant a$, and the value of $\varkappa$, if ψ_0 vanishes only for $z=0$ and $z=a$.

2. An E_{mn}-wave of frequency $\omega/2\pi$ is propagated along a perfectly conducting wave-guide whose sides are the planes $x=0$, $x=a$, $y=0$, $y=b$. Show that the electric field component along the axis of the guide is of the form

$$E_z=A\sin\left(\frac{m\pi x}{a}\right)\sin\left(\frac{n\pi y}{b}\right)$$

where

$$\beta^2=\left(\frac{\omega}{c}\right)^2-\pi^2\left(\frac{m^2}{a^2}+\frac{n^2}{b^2}\right)$$

Determine the other field components, and obtain an expression for the energy-flow along the guide.

3. An electromagnetic E-wave of frequency $\omega/2\pi$ is propagated along a rectangular wave guide having perfectly conducting walls $x=0$, $x=a$, $y=0$, $y=b$. Obtain the field components in the form

$$E_x=-\frac{\mathrm{i}\beta A}{v^2}\frac{m\pi}{a}\cos\left(\frac{m\pi x}{a}\right)\sin\left(\frac{n\pi y}{b}\right);\qquad B_x=\frac{A\mathrm{i}\omega K}{v^2c^2}\frac{n\pi}{b}\sin\left(\frac{m\pi x}{a}\right)\cos\left(\frac{n\pi y}{b}\right);$$

$$E_y=-\frac{\mathrm{i}\beta A}{v^2}\frac{n\pi}{b}\sin\left(\frac{m\pi x}{a}\right)\cos\left(\frac{n\pi y}{b}\right);\qquad B_y=-\frac{A\mathrm{i}\omega K}{v^2c^2}\frac{m\pi}{a}\cos\left(\frac{m\pi x}{a}\right)\sin\left(\frac{n\pi y}{b}\right);$$

$$E_z=A\sin\left(\frac{m\pi x}{a}\right)\sin\left(\frac{n\pi y}{b}\right);\qquad B_z=0,$$

where K is the dielectric constant of the medium filling the guide, the permeability is unity,

$$\beta^2=\frac{\omega^2}{v^2}-\nu^2,$$

$$\nu^2=\left(\frac{m\pi}{a}\right)^2+\left(\frac{n\pi}{b}\right)^2,$$

m and n are integers and v is the velocity of propagation in the unbounded medium.

Show that at all points in the field, the electric and magnetic fields are perpendicular and that there is a phase difference of $\pi/2$ between the magnetic vector and the longitudinal resolute of the electric vector.

4. Prove that a wave whose field components are

$$E_x=A\sin\left(\frac{n\pi y}{b}\right)\exp\{\mathrm{i}(\omega t-\beta z)\},$$

$$H_y=B\sin\left(\frac{n\pi y}{b}\right)\exp\{\mathrm{i}(\omega t-\beta z)\},$$

$$H_z=C\cos\left(\frac{n\pi y}{b}\right)\exp\{\mathrm{i}(\omega t-\beta z)\},$$

where n is an integer, may be propagated in the direction of Oz between two perfectly conducting planes $y = 0$ and $y = b$, in a medium of dielectric constant K and permeability μ. Give the equation satisfied by β and find the ratio $A : B : C$.

If λ is the wavelength of these waves and λ_0 that for waves of the same frequency in an unbounded medium, show that

$$\frac{1}{\lambda^2} = \frac{1}{\lambda_0^2} - \frac{n^2}{4b^2}.$$

Under what conditions could this field exist in a rectangular waveguide?

5. Show that Maxwell's equations for an isotropic homogeneous non-conducting medium of permeability μ and permittivity ε can be satisfied by taking

$$\boldsymbol{E} = \text{real part of } \mathbf{curl\ curl}\,(\psi \boldsymbol{k}),$$

$$\boldsymbol{B} = \text{real part of } \frac{\partial}{\partial t}\,\mathbf{curl}\,(\psi \boldsymbol{k})$$

provided that ψ is a solution of the wave equation $\nabla^2\psi = \mu\varepsilon\partial^2\psi/\partial t^2$.

Taking $\psi = XY \exp\{i(\omega t - \beta z)\}$, where X is a function of x only and Y is a function of y only, obtain the components of $\boldsymbol{E}$, $\boldsymbol{B}$ for a field in a rectangular waveguide bounded by perfectly conducting planes $x = 0$, $x = a$, $y = 0$, $y = b$, and find the condition that this field should be propagated without attenuation.

11.8 The transmission line

Like a waveguide, a transmission line is an arrangement of conductors which "guide" the propagation of an oscillatory electromagnetic field. Whereas the waveguide is a hollow conductor with the field propagated down the inside, a transmission line consists, usually, of two parallel conductors and the field is established between them. One kind of line is exemplified in telegraph or telephone wires, or power cables, where the conductors are outside one another. Another kind is the coaxial cable where one conductor completely encloses the other and the field is established in the space between them.

We shall discuss chiefly the simplest case of two straight conductors, having infinite electrical conductivity, which are embedded in a uniform medium. This medium may have a finite electrical conductivity corresponding, for example, to a submarine cable. In practice the medium surrounding the conductors is not uniform, e.g. there are layers of special insulation surrounding the wires, nor is the conductivity of the wires infinite.

We saw that, unless the section of a waveguide was a doubly- (or multiply-) connected region, only TM- or TE-type waves were possible; waves of TEM-type were impossible, and the "cut-off" frequency was due to the reflections which took place from the walls of the tube. Since the transmission line field is established *outside* the conductors these reflections (from the outer walls) do not take place. Consequently, only TEM-type waves can occur; in this case TM- and TE-type waves cannot be established with perfectly conducting boundaries.

The "field outlook" regards the space where the field is established as the primary seat of the phenomena of electromagnetism. Hence, corresponding to this field there are distributions of current, charge and voltage on the conductors of the transmission line. In most practical applications these currents and voltages—the signals transmitted—are of more importance than the field strengths. Also the quantities we consider all vary harmonically with time and so, effectively, the conductors carry alternating currents and voltages. So it is natural to describe the behaviour of the line in terms of impedances, which relate the currents and voltages, as in alternating current theory.

The arrangement we consider is made up of one or more, usually two, long, straight conductors with uniform cross-section perpendicular to the z-direction. We look for fields which correspond to the propagation of waves in the z-direction by the same assumptions as in § 11.7 and use expressions (11.122) for the field strengths. The analysis is identical as far as eqn. (11.131). Because the field now extends to infinity we must impose conditions to be satisfied at infinity, as well as on the conducting boundaries, before we can determine *unique* solutions to the field equations. These conditions are usually known as the "Sommerfeld radiation conditions". We cannot give a thorough discussion of these here, and content ourselves with the following account which draws an analogy with the conditions for uniqueness in electrostatics. The "standard" boundary conditions applied there for a field in two dimensions (and uniform in the z-direction) were

$$V = O(\ln r), \quad |\boldsymbol{E}| = O(1/r), \quad r \to \infty .$$

In physical terms this means that the potential and field correspond to the field of a line charge through the origin; this is the simplest cylindrically symmetric solution of Laplace's equation.

In considering TE- and TM-waves we found that the functions ϕ, ψ had to satisfy Helmholtz's equation for two dimensions, viz.

$$\frac{\partial^2 \phi}{\partial x^2} + \frac{\partial^2 \phi}{\partial y^2} + \nu^2 \phi = 0.$$

The "radiation conditions" require that at infinity ϕ, or ψ, should correspond to the simplest (cylindrically) symmetric wave propagated from a source lying along the z-axis. The conditions for this are

$$\phi = O(1/r^{1/2}), \quad r \to \infty ; \qquad \lim_{r \to \infty} r^{1/2} \left(\frac{\partial \phi}{\partial r} + i\nu\phi \right) = 0. \qquad (11.137)$$

When these conditions are satisfied at infinity, Helmholtz's equation has a unique solution if ϕ and/or $\partial\phi/\partial n$ is specified on finite boundaries. (The conditions are sufficient for uniqueness; this does not imply that they are also necessary.) Equation (11.137) replaces the qualitative statement about the absence of reflections. The consequence of applying (11.137) is that TE- and TM-waves do not exist outside perfectly conducting transmission lines.

Example. We illustrate the above by considering the case of a perfectly conducting cylinder of radius *a*.

Because of the complete cylindrical symmetry we look for an axially symmetric solution of

$$\frac{\partial^2\phi}{\partial x^2}+\frac{\partial^2\phi}{\partial y^2}+\nu^2\phi = 0$$

which vanishes for $r = a$. In cylindrical polars this becomes, because ϕ is independent of θ,

$$\frac{d^2\phi}{dr^2}+\frac{1}{r}\frac{d\phi}{dr}+\nu^2\phi = 0,$$

which has the general solution

$$\phi = AJ_0(\nu r)+BY_0(\nu r),$$

where $J_0(\nu r)$ and $Y_0(\nu r)$ are Bessel functions of the first and second kinds of order zero. To investigate the behaviour of ϕ at infinity we use the asymptotic forms

$$J_0(\nu r) \approx \frac{\sin \nu r+\cos \nu r}{\sqrt{(\pi\nu r)}}, \quad Y_0(\nu r) \approx \frac{\sin \nu r-\cos \nu r}{\sqrt{(\pi\nu r)}}$$

for large values of *r*.

It is clear that the first of conditions (11.137) is satisfied for

$$|r^{1/2}\phi(r)| \approx |(A+B)\sin \nu r+(A-B)\cos \nu r|/\sqrt{(\pi\nu)} < K.$$

Also

$$r^{1/2}\left(\frac{\partial\phi}{\partial r}+i\phi\right) = \frac{(A+iA+B-iB)\nu\cos \nu r+(-A+iA+B+iB)\nu\sin \nu r}{\sqrt{(\pi\nu)}} - \frac{(A+B)\sin \nu r+(A-B)\cos \nu r}{2r\sqrt{(\pi\nu)}}.$$

If the second of conditions (11.137) is to be satisfied, the coefficients of $\cos \nu r$ and $\sin \nu r$ in the first fraction must both vanish, i.e.

$$A(1+i)+B(1-i) = 0, \quad A(-1+i)+B(1+i) = 0.$$

Therefore
$$B = -iA$$
and
$$\phi(r) = A\{J_0(\nu r)-iY_0(\nu r)\} = AH_0^{(2)}(\nu r).$$

This particular combination of J_0 and Y_0 is known as a Hankel function, $H_0^{(2)}(\nu r)$, which has the property that it does not vanish for any real value of the argument. Hence we can only satisfy the boundary condition $\phi(a) = 0$ by taking $A = 0$. Therefore, as long as

$\nu \neq 0$, there is no solution corresponding to TE- or TM-waves. (A similar argument applies to solutions which depend on the polar angle θ. Again there are no TE- or TM-solutions.)

Since only TEM-waves are possible the field vectors satisfy eqns. (11.122), with $E_z = 0 = B_z$. Hence

$$\frac{\partial e_x}{\partial x}+\frac{\partial e_y}{\partial y}=0, \quad \frac{\partial b_x}{\partial x}+\frac{\partial b_y}{\partial y}=0,$$
$$\frac{\partial e_x}{\partial y}-\frac{\partial e_y}{\partial x}=0, \quad \frac{\partial b_x}{\partial y}-\frac{\partial b_y}{\partial x}=0, \tag{11.138}$$
$$e_x=-(\omega/\gamma)b_y, \quad e_y=(\omega/\gamma)b_x, \quad b_x=(\omega\mu\varepsilon/\gamma)e_y, \quad b_y=-(\omega\mu\varepsilon/\gamma)e_x,$$

so that

$$\gamma^2=\omega^2\mu\varepsilon,$$

(i.e. $\nu = 0$). We use μ, ε, rather than μ_0, ε_0, so that we can include in our discussion the case of conductors embedded in a medium, instead of a vacuum, and, by using a complex value for the permittivity $\varepsilon' = \varepsilon - i\sigma/\omega$, we can also include the case of a surrounding medium with conducting properties. When ε is real, the velocity of propagation of the waves is $v = \omega/\gamma = 1/\sqrt{(\mu\varepsilon)}$.

The solution of eqns. (11.138) is obtained by introducing a potential function $\mathcal{V}$ as follows:

$$e_x=-\frac{\partial\mathcal{V}}{\partial x}, \quad e_y=-\frac{\partial\mathcal{V}}{\partial y}, \quad b_x=\frac{\gamma}{\omega}\frac{\partial\mathcal{V}}{\partial y}, \quad b_y=-\frac{\gamma}{\omega}\frac{\partial\mathcal{V}}{\partial x}, \tag{11.139}$$

where

$$\frac{\partial^2\mathcal{V}}{\partial x^2}+\frac{\partial^2\mathcal{V}}{\partial y^2}=0.$$

In addition $\mathcal{V}$ must be constant on the boundaries, or $\partial\mathcal{V}/\partial n$ must be specified. At infinity $\mathcal{V}$ must satisfy the radiation conditions (11.137).

We notice that $\mathcal{V}$ is the potential function of a two dimensional electrostatic field. Since the first of conditions (11.137) prevents $\mathcal{V}$ from containing a term $\ln r$, the net charge residing on the finite boundaries of this electrostatic field must be zero. Because the charge on a single isolated conductor in a field must be all of one sign (see Vol. 1, p. 80) it follows that no wave, of the kind we consider here, can be propagated along a *single perfectly conducting* guide. Therefore, in what follows we shall consider a transmission line consisting of two conductors. [The assumption that we make by adopting the forms (11.122) as our starting-point is that the harmonic wave extends uniformly along the whole (infinite) length of the line. If a charge, for example, is situated on part of the line a disturbance is propagated, both ways, along the line with velocity v, but this disturbance cannot be represented by a wave of the kind considered here with a *single* frequency. If the line is not a perfect conductor, then a wave can be propagated along it, but this wave is not of the TEM-type.]

With the addition of the factor $\exp\{i(\omega t-\gamma z)\}$ the expressions in (11.139) give the instantaneous values of the field quantities. Hence the "static" field given by $\mathcal{V}$ gives the amplitude, in any transverse plane, of the oscillatory quantities which constitute the complete field. We now investigate what charges, currents and voltages on the conductors of the line correspond to these fields.

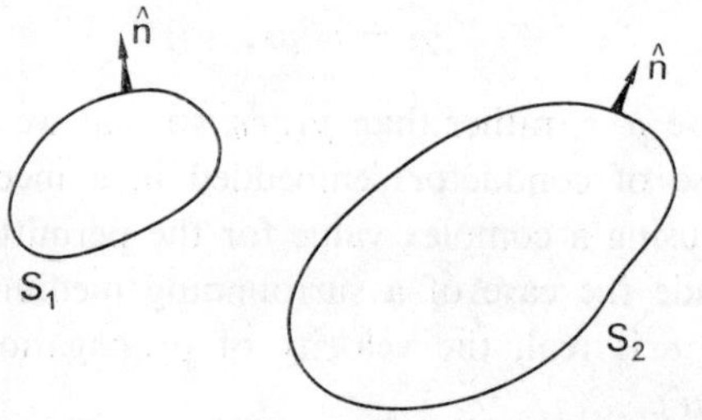

FIG. 11.13

We consider two curves S_1 and S_2 in the xy-plane (Fig. 11.13) which are the profiles of the conductors. The static field $\mathcal{V}$ corresponds to charges Q_1, Q_2 on unit length of these conductors, where

$$Q_1+Q_2 = 0 \tag{11.140}$$

and

$$Q_1 = -\varepsilon \int_{S_1} \frac{\partial \mathcal{V}}{\partial n}\, ds, \quad Q_2 = -\varepsilon \int_{S_2} \frac{\partial \mathcal{V}}{\partial n}\, ds. \tag{11.141}$$

Also the surface current density on the boundary corresponding to the magnetic field is given by $\hat{\boldsymbol{n}}\times\boldsymbol{H} = (\hat{\boldsymbol{n}}\times\boldsymbol{B})/\mu$. This corresponds to total currents $\mathcal{I}_1$ and $\mathcal{I}_2$ flowing in the z-direction along each conductor given by

$$\mathcal{I}_1 = -\frac{\gamma}{\omega\mu} \int_{S_1} \frac{\partial \mathcal{V}}{\partial n}\, ds, \quad \mathcal{I}_2 = -\frac{\gamma}{\omega\mu} \int_{S_2} \frac{\partial \mathcal{V}}{\partial n}\, ds, \tag{11.142}$$

since $(\boldsymbol{n}\times\boldsymbol{B})/\mu = -(\gamma/\omega\mu)(\partial\mathcal{V}/\partial n)\boldsymbol{k}$. Hence we see that $\mathcal{I}_1 = -\mathcal{I}_2$ and the current flowing along one conductor returns along the other. If we denote these currents and charges by $\pm\mathcal{I}$, $\pm\mathcal{Q}$, we see that

$$\mathcal{I} = \gamma\mathcal{Q}/(\omega\mu\varepsilon) = \mathcal{Q}/\sqrt{(\mu\varepsilon)}. \tag{11.143}$$

The conductors in the "static" field are at potentials $\mathcal{V}_1$ and $\mathcal{V}_2$ where $\mathcal{V} = \mathcal{V}_1 - \mathcal{V}_2$, and so the capacitance C of unit length of the two conductors is given by

$$C(\mathcal{V}_1 - \mathcal{V}_2) = \mathcal{Q} = C\mathcal{V} \tag{11.144}$$

and C may be calculated from the potential function but depends only on the shape and relative dispositions of the curves S_1, S_2. We also define the inductance per unit length, L, in terms of the magnetic energy

$$\frac{1}{2}L\mathcal{I}^2 = \frac{1}{2\mu}\iint \boldsymbol{B}^2\,\mathrm{d}x\,\mathrm{d}y,$$

where the integral is taken through the field outside the conductors. Then

$$\begin{aligned}\frac{1}{2}L\mathcal{I}^2 &= \frac{\gamma^2}{2\omega^2\mu}\iint\left\{\left(\frac{\partial\mathcal{V}}{\partial x}\right)^2 + \left(\frac{\partial\mathcal{V}}{\partial y}\right)^2\right\}\mathrm{d}x\,\mathrm{d}y\\ &= -\frac{\gamma^2}{2\omega^2\mu}\iint \mathcal{V}\frac{\partial\mathcal{V}}{\partial n}\,\mathrm{d}s,\end{aligned}$$

where we have used Green's theorem remembering that $\partial/\partial n$ is here taken into the field region. The contribution from infinity is zero. Therefore

$$\frac{1}{2}L\mathcal{I}^2 = \frac{\gamma^2}{2\omega^2\mu\varepsilon}(\mathcal{V}_1\mathcal{Q}_1 + \mathcal{V}_2\mathcal{Q}_2) = \frac{\gamma^2\mathcal{V}\mathcal{Q}}{2\omega^2\mu\varepsilon} = \frac{1}{2}\mathcal{V}\mathcal{Q}. \tag{11.145}$$

These results mean that at any point on the length of the conductors the current strength is given by (the real part of)

$$I = \mathcal{I}\{\exp \mathrm{i}(\omega t - \gamma z)\}$$

and the potential difference between the conductors at this point is (the real part of)

$$V = \mathcal{V}\exp\{\mathrm{i}(\omega t - \gamma z)\}.$$

These quantities are found in a system with a capacitance C and inductance L per unit length. From eqn. (11.144) we have $\mathcal{Q} = C\mathcal{V}$, and, from (11.145),

$$L\mathcal{I} = \mathcal{Q}\mathcal{V}/\mathcal{I} = \sqrt{(\mu\varepsilon)}\mathcal{V} = (\gamma/\omega)\mathcal{V}.$$

Therefore $\dfrac{\partial V}{\partial z} = -\mathrm{i}\gamma\mathcal{V}\exp\{\mathrm{i}(\omega t - \gamma z)\} = -\mathrm{i}\omega L\mathcal{I}\exp\{\mathrm{i}(\omega t - \gamma z)\} = -\mathrm{i}\omega LI.$

(11.146)

Also, since $\mathcal{I} = (\omega/\gamma)\mathcal{Q} = (\omega/\gamma)C\mathcal{V}$,

$$\frac{\partial I}{\partial z} = -i\gamma\left(\frac{\omega}{\gamma}\right)CV = -i\omega CV. \tag{11.147}$$

If we remember that in this case $\partial/\partial t \equiv i\omega$ these results are equivalent to

$$\frac{\partial V}{\partial z}+L\frac{\partial I}{\partial t} = 0, \quad \frac{\partial I}{\partial z}+C\frac{\partial V}{\partial t} = 0, \tag{11.148}$$

which lead to

$$\frac{\partial^2 I}{\partial z^2} = LC\frac{\partial^2 I}{\partial t^2}, \quad \frac{\partial^2 V}{\partial z^2} = LC\frac{\partial^2 V}{\partial t^2}. \tag{11.149}$$

These latter equations, (11.149), are known as the "equations of telegraphy" for a loss-free transmission line. (We can write $i\omega I = \partial I/\partial t$ in general because an arbitrary periodic function I of the time can, by Fourier's theorem, be written as the sum of a number of different frequencies, to each of which this relation applies.) Since V, I are alternating quantities, we define the characteristic impedance of the line Z_C by

$$Z_C = V/I = \mathcal{V}/\mathcal{I} = \frac{\omega}{\gamma C} = \frac{\sqrt{(\mu\varepsilon)}}{C} = \frac{L}{\sqrt{(\mu\varepsilon)}}. \tag{11.150}$$

In this case the current I and voltage V are in phase.

We consider now what modifications become necessary if the medium surrounding the conductors is a conducting medium (with a conductivity small compared with that of the line itself). In practice this means that the current entering an element of the line divides into a component which goes on down the line and another, small, component (the "shunt" current) which "leaks" across the medium to the other line conductor. The modification we make is to use a complex permittivity $\varepsilon' = \varepsilon - i\sigma/\omega$. Since $\gamma[= (\mu\varepsilon)^{1/2}\{1-i\sigma/(\varepsilon\omega)\}^{1/2}]$ is complex, the imaginary part of γ leads to attenuation of all quantities, including V, I along the line.

We consider unit length of line and the field enclosed between two planes z = constant at unit distance apart. The mean rate of dissipation of heat in this part of the field is given by (the real part of) $\frac{1}{2}\sigma \boldsymbol{E}\cdot\boldsymbol{E}^*$. If we regard this heat as produced by the passage of the "leak" current flowing between the conductors at a potential difference $\mathcal{V}$ the expression for this is $\frac{1}{2}G\mathcal{V}\mathcal{V}^*$, where G is the conductance, i.e. the reciprocal of the resistance, between unit length of the conductors of the line. Then

$$\frac{1}{2}G\mathcal{V}\mathcal{V}^* = \frac{1}{2}\sigma\iint \boldsymbol{E}\cdot\boldsymbol{E}^*\,\mathrm{d}x\,\mathrm{d}y = \frac{1}{2}\sigma\iint\left\{\frac{\partial\mathcal{V}}{\partial x}\frac{\partial\mathcal{V}^*}{\partial x}+\frac{\partial\mathcal{V}}{\partial y}\frac{\partial\mathcal{V}^*}{\partial y}\right\}\mathrm{d}x\,\mathrm{d}y$$

$$= -\frac{1}{2}\sigma\int_{S_1+S_2}\mathcal{V}\frac{\partial\mathcal{V}^*}{\partial n}\,\mathrm{d}s = \frac{1}{2}\sigma\frac{\mathcal{V}Q^*}{\varepsilon}.$$

Therefore $G\mathcal{V} = \sigma Q/\varepsilon$, i.e. $G = \sigma C/\varepsilon$.

If we use ε', instead of ε, in the expression for the capacitance, we obtain a complex value

$$C' = -\frac{\varepsilon'}{\mathcal{V}}\int\frac{\partial\mathcal{V}}{\partial n}\,\mathrm{d}s = -\left(1-\frac{\mathrm{i}\sigma}{\omega\varepsilon}\right)\frac{\varepsilon}{\mathcal{V}}\int\frac{\partial\mathcal{V}}{\partial n}\,\mathrm{d}s = C-\frac{\mathrm{i}\sigma C}{\omega\varepsilon}.$$

Hence

$$C' = C-\mathrm{i}G/\omega.$$

Hence the modifications we have to make to the results already obtained are:

from (11.146) $\dfrac{\partial V}{\partial z} = -\mathrm{i}\omega LI$, (no change); (11.151)

from (11.147) $\dfrac{\partial I}{\partial z} = -\mathrm{i}\omega C'V$ becomes $\dfrac{\partial I}{\partial z} = -(\mathrm{i}\omega C+G)V$; (11.152)

from (11.148) $\dfrac{\partial V}{\partial z}+L\dfrac{\partial I}{\partial t} = 0,\quad \dfrac{\partial I}{\partial z}+C\dfrac{\partial V}{\partial t}+GV = 0;$ (11.153)

and from (11.150)

$$L(C-\mathrm{i}G/\omega) = \mu(\varepsilon-\mathrm{i}\sigma/\omega). \tag{11.154}$$

Finally, we consider line conductors which have a large, but not infinite, conductivity. When this is the case there must be a component of $\boldsymbol{E}$ in the direction of the line because, in general, there is a current flowing in the z-direction through the resistance of the line. Hence $E_z \neq 0$, and the field is no longer of the TEM-type. We assume, for lines made of good conductors, that the TEM-field gives a sufficiently close approximation to the field outside these conductors, but inside these conductors there is the necessary longitudinal component E_z. The alternating current $\mathcal{J}\exp\{\mathrm{i}(\omega t-\gamma z)\}$ is not now exactly a surface current, but it penetrates into the conductor through the skin effect. This has two consequences. First in addition to the variation of voltage in the z-direction given by $\partial V/\partial z = -\mathrm{i}\gamma\mathcal{V}\exp\{\mathrm{i}(\omega t-\gamma z)\}$ we must have a variation $\mathcal{J}R_i\,\delta z$ in a length

δz due to Ohmic resistance, the "internal" resistance of the line R_i per unit length. Hence the voltage δV between two points separated by a line-distance δz is given by

$$\delta V = -i\gamma \mathcal{V}\, \delta z \exp\{i(\omega t - \gamma z)\} - IR_i\, \delta z.$$

Therefore

$$\frac{\partial V}{\partial z} = -IR_i - i\gamma \mathcal{V} \exp\{i(\omega t - \gamma z)\}.$$

The second consequence is the addition of an internal inductance L to L. The magnetic energy in unit thickness of the field is (the real part of) $1/(4\mu) \iint \boldsymbol{B}\cdot\boldsymbol{B}^*\, dx\, dy$, where the integral must now include the cross-section of the line-conductors where the current is no longer strictly zero. We therefore write

$$\frac{1}{4\mu}\iint \boldsymbol{B}\cdot\boldsymbol{B}^*\, dx\, dy = \frac{1}{4\mu}\iint_{\infty - S_1 - S_2} \boldsymbol{B}\cdot\boldsymbol{B}^*\, dx\, dy + \frac{1}{4\mu}\iint_{S_1+S_2} \boldsymbol{B}\cdot\boldsymbol{B}^*\, dx\, dy$$
$$= \frac{1}{2} L\mathcal{I}\mathcal{I}^* + \frac{1}{2} L_i\mathcal{I}\mathcal{I}^*.$$

The first integral is taken outside the conductors $(\infty - S_1 - S_2)$ and is the same as in eqn. (11.145); the second integral over (S_1+S_2) is the internal integration. Hence we replace L by $L+L_i$, where L_i is the internal inductance of the wires in the line. The further modifications necessary in eqns. (11.151–153) now give

$$\frac{\partial V}{\partial z} = -I[R_i + i\omega(L+L_i)], \quad \frac{\partial I}{\partial z} = -(i\omega C + G)V. \tag{11.155}$$

If, instead of the harmonic time variation, we consider any variation, we have the results

$$\frac{\partial V}{\partial z} + (L+L_i)\frac{\partial I}{\partial t} + R_i I = 0, \quad \frac{\partial I}{\partial z} + C\frac{\partial V}{\partial t} + GV = 0. \tag{11.156}$$

Elimination of V, or I, leads to the final version of the "equation of telegraphy".

$$\frac{\partial^2 I}{\partial z^2} - C(L+L_i)\frac{\partial^2 I}{\partial t^2} - \{CR_i + G(L+L_i)\}\frac{\partial I}{\partial t} - R_i GI = 0, \tag{11.157}$$

which is also satisfied by V.

Example. For our subsequent use we consider here a network of impedances which is made up of a sequence of elements $A_rA_{r+2}B_{r+1}B_r$ connected end to end forming two conducting lines $A_1A_2 \ldots A_rA_{r+1}A_{r+2} \ldots$ and $B_1B_2 \ldots B_rB_{r+1}B_{r+2} \ldots$ as in Fig. 11.14.

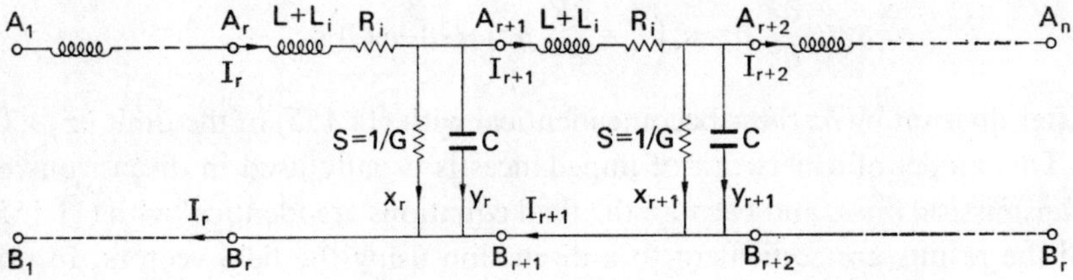

FIG. 11.14

We denote the current entering each element at A_r, and leaving at B_r by I_r, at A_{r+1} by I_{r+1}, and so on; similarly the potential difference between A_r, B_r is V_r, between A_{r+1}, B_{r+1} is V_{r+1}, and so on. The equations for the network, when the time variation is given by the factor $e^{i\omega t}$, are

$$V_r - V_{r+1} = I_r[R_i + i\omega(L+L_i)], \tag{1}$$

$$V_{,+1} = Sx_r = x_r/G, \qquad V_{r+1} = y_r/(i\omega C),$$

$$I_r = I_{r+1} + x_r + y_r,$$

where x_r and y_r are the currents "leaking" through the "shunts" S and C. The latter equations give

$$I_r - I_{r+1} = x_r + y_r = V_{r+1}(G + i\omega C). \tag{2}$$

The results (1) and (2) of the above example show an obvious (and contrived) resemblance to the equations of the transmission line. We can make the resemblance complete if we regard the element $A_rA_{r+1}B_{r+1}B_r$ as a model of part of the transmission line of length δz. Then we make the following identifications:

$$I_r = I, \quad I_{r+1} = I + \frac{\partial I}{\partial z}\,\delta z, \quad V_r = V, \quad V_{r+1} = V + \frac{\partial V}{\partial z}\,\delta z,$$

and the impedances are replaced as follows:

$$R_i \quad \text{by} \quad R_i\,\delta z, \qquad L+L_i \quad \text{by} \quad (L+L_i)\,\delta z,$$

$$\frac{1}{S} = G \quad \text{by} \quad G\,\delta z, \qquad C \quad \text{by} \quad C\,\delta z.$$

In the latter two identifications we are using the rules for connecting resistances in parallel, and capacitors in parallel. Then eqns. (1) and (2) of the above example become

$$\frac{\partial V}{\partial z}\,\delta z = I[R_i + i\omega(L+L_i)]\,\delta z,$$

$$\frac{\partial I}{\partial z}\,\delta z = \left(V+\frac{\partial V}{\partial z}\,\delta z\right)(G+i\omega C)\,\delta z.$$

After division by δz these become identical with (11.155) in the limit $\delta z \to 0$.

This model of a network of impedances is usually used in discussions of transmission lines, and because the final equations are identical with (11.155) all the results are equivalent to a discussion using the field vectors. In this model the conductor $B_1 \ldots B_r \ldots B_n$ is usually taken to be the earth, and the line consists of the *single* conductor $A_1 \ldots A_r \ldots A_n$. This differs from our original specification of *two* conductors. The discrepancy can be avoided if we regard the two conductors as separated by an infinite conducting plane at zero potential. The second line is then the image of the line of the first conductor in this plane. This plane is taken to be the earth conductor in the model.

We now complete our discussion of the transmission line starting from the equations (11.156–157) and pay little attention to the field strengths; we shall use a model of the type we have just used to make any modifications for special cases. So far we have considered an infinite line, but now we shall consider a "long" line, i.e. one whose total length in the z-direction is very large compared with the separation of the conductors, or their dimensions; a "signal" is applied at one end, consisting of a varying voltage or current which is transmitted down the line and at the other end the two conductors are connected together through an impedance representing the receiving apparatus. The mathematical discussion consists of finding a solution of eqns. (11.156–157) subject to boundary conditions at $z=0$ and $z=l$ (or ∞).

Because of the resistance in the conductors the signal suffers attenuation as it is transmitted down the line; the important practical case is that in which attenuation occurs without distortion. If the attenuation depends on the frequency, a given signal, which contains many different frequencies, will have these different components attenuated to different degrees when it reaches the end of the line, and so suffers distortion. Similarly distortion occurs if the different frequencies are transmitted with velocities which depend on the frequency. We shall see that by adjusting the values of the parameters R_i, L, L_i, C, G a line can be made distortionless. Such a line can have amplifiers inserted into it at suitable points along its length to overcome the attenuation, and a clear signal can then be transmitted over long distances.

To simplify the working we drop the suffix i (for "internal") and combine $L+L_i$ into a single term L to denote the total inductance of unit length of the line. The equations governing the behaviour of the line are therefore

$$\frac{\partial V}{\partial z}+L\frac{\partial I}{\partial t}+RI=0,\qquad \frac{\partial I}{\partial z}+C\frac{\partial V}{\partial t}+GV=0. \qquad (11.158\,\text{a, b})$$

For a signal of frequency $\omega/2\pi$ these become

$$\frac{\partial V}{\partial z}+(R+\mathrm{i}\omega L)I=0,\qquad \frac{\partial I}{\partial z}+(G+\mathrm{i}\omega C)V=0. \qquad (11.159\,\text{a, b})$$

A wave given by $V=\mathcal{V}\exp\{\mathrm{i}(\omega t\pm\gamma z)$, $I=\mathcal{I}\exp\{\mathrm{i}(\omega t\pm\gamma z)$ is propagated along the line if

$$\pm\mathrm{i}\gamma V+(R+\mathrm{i}\omega L)I=0,\qquad \pm\mathrm{i}\gamma I+(G+\mathrm{i}\omega C)V=0,$$

$$\therefore\quad \gamma^2=-(R+\mathrm{i}\omega L)(G+\mathrm{i}\omega C)=(\omega^2LC-RG)-\mathrm{i}\omega(LG+RC).$$

Since $LG+RC\geqslant 0$ we write $\gamma=\alpha-\mathrm{i}\beta$ where

$$\alpha^2-\beta^2=\omega^2LC-RG,\qquad 2\alpha\beta=\omega(LG+RC), \qquad (11.160)$$

so that α,β are either both positive or both negative, corresponding to the two cases $+\gamma$, $-\gamma$. When $\alpha,\beta>0$ the wave is

$$V=\mathcal{V}\mathrm{e}^{-\beta z}\exp\{\mathrm{i}(\omega t-\alpha z)\},\qquad I=\mathcal{I}\mathrm{e}^{-\beta z}\exp\{\mathrm{i}(\omega t-\alpha z)\}.$$

The velocity of propagation of the wave along the line is ω/α, and the factor $\mathrm{e}^{-\beta z}$ gives the attenuation; the positive signs for α,β correspond to propagation in the positive z-direction with attenuation; and the negative signs for α,β correspond to propagation in the negative direction, also with attenuation.

If the quantities α,β are independent of ω, then all frequencies are propagated with the same speed and suffer the same attenuation. In this case the line is distortionless. If we eliminate α from eqns. (11.160), we find

$$\omega^2(LG+RC)^2-4\beta^2(\omega^2LC-RG)-4\beta^4=0.$$

The value of β is independent of ω if the coefficient of ω^2 vanishes, i.e. if

$$4\beta^2LC=(LG+RC)^2,\quad 4\beta^2RG-4\beta^4=0,\quad \text{i.e.}\quad \beta^2=GR.$$

Therefore $\quad 4RGLC=(LG+RC)^2,\quad$ i.e. $\quad (LG-RC)^2=0.$

Hence for a distortionless line we must have

$$\frac{L}{C} = \frac{R}{G}, \quad \beta^2 = RG, \quad \alpha^2 = \omega^2 LC, \tag{11.161}$$

and the velocity of propagation is $(LC)^{-1/2}$.

The general solution of the eqns. (11.159) corresponding to a harmonic wave is

$$I = (Pe^{-i\gamma z}+Qe^{i\gamma z})e^{i\omega t}, \quad V = i\gamma(-Pe^{-i\gamma z}+Qe^{i\gamma z})e^{i\omega t}/(G+i\omega C),$$

and corresponds to a wave in each direction. For a single wave travelling in (say) the direction $z > 0$, at any point we have (with $Q = 0$)

$$I = P \exp\{i(\omega t-\gamma z)\}, \quad V = \frac{-i\gamma P}{G+i\omega C} \exp\{i(\omega t-\gamma z)\}.$$

Therefore

$$Z_C = \frac{V}{I} = \frac{-i\gamma}{G+i\omega C} = \left(\frac{R+i\omega L}{G+i\omega C}\right)^{1/2}. \tag{11.162}$$

The quantity Z_C is the *characteristic impedance* of the line.

Thus far we have not considered the effect of the ends of the line. We suppose that the signal put into the line at A, $z = 0$, is given by current I_A and voltage V_A. Then the impedance "seen by the transmitter" is $Z_A = V_A/I_A$, where

$$I_A = (P+Q)e^{i\omega t}, \quad V_A = Z_C(P-Q)e^{i\omega t}.$$

At the end B, where $z = l$, we suppose that there is an impedance Z_l (the load) connecting the two conductors of the line. For $z = l$ we have

$$I_B = (Pe^{-i\gamma l}+Qe^{i\gamma l})e^{i\omega t}, \quad V_B = Z_C(Pe^{-i\gamma l}-Qe^{i\gamma l})e^{i\omega t}.$$

Then

$$\frac{V_B}{I_B} = Z_B = Z_l = Z_C \frac{Q\,e^{i\gamma l}-P\,e^{-i\gamma l}}{Q\,e^{i\gamma l}+P\,e^{-i\gamma l}}.$$

From these equations we deduce that

$$\frac{Q}{P} = \frac{Z_C-Z_l}{Z_C+Z_l}e^{-2i\gamma l}, \quad \frac{Z_A}{Z_C} = \frac{Z_l+iZ_C \tan \gamma l}{Z_C+iZ_l \tan \gamma l}. \tag{11.163}$$

We can draw certain conclusions from these results. In general we see that every signal must give rise to a reflection from the far end ($Q \neq 0$, in general). There are three particular results:

1. If $Z_l = Z_C$, $Q = 0$ and there is no reflection.
   The line is then *matched* and $Z_A = Z_C$.

2. When the line is on *open-circuit* Z_l is infinite, $Z_A = -iZ_C \cot \gamma l$, and $Q = -P e^{-2i\gamma l}$, so that there is a phase difference $\pi - 2\gamma l$ between the signal at A and its reflection.
3. When the line is *short-circuited* $Z_l = 0$, then

$$Z_A = iZ_C \tan \gamma l, \quad Q = P e^{-2i\gamma l},$$

If there are no (heat) losses $R = 0$ and $G = 0$, so that $\gamma^2 = \omega^2 LC$ and γ is real, and there is no attenuation. In general, the solution of problems concerning transmission lines requires ideas and techniques very closely similar to those required for waves on strings, both finite and infinite in length.

Example. In a transmission line $A_0A_1A_2 \ldots A_{n+1}$ each of the elements $A_1A_2, \ldots, A_{n-1}A_n$ has inductance L, and each of the junctions $A_1, A_2, \ldots, A_n$ is connected to earth through a capacitor of capacitance C. Resistance is negligible, and the impedance of the elements A_0A_1 and A_nA_{n+1} is zero. A (complex) alternating current x_0 is fed in at the terminal A_0, and x_r is the current along A_rA_{r+1}; show that, provided $LC\omega^2 < 4$,

$$x_r - 2x_{r+1} \cos \theta + x_{r+2} = 0,$$

where

$$\cos \theta = 1 - \tfrac{1}{2} LC\omega^2.$$

Write down the general solution of this equation.

Show that if the terminal A_{n+1} is connected to earth through a receiver of resistance R and inductance $\frac{1}{2}L$, and if $RC\omega = \pm \sin \theta$ then the current x_n through the receiver is $x_0 e^{\mp in\theta}$.

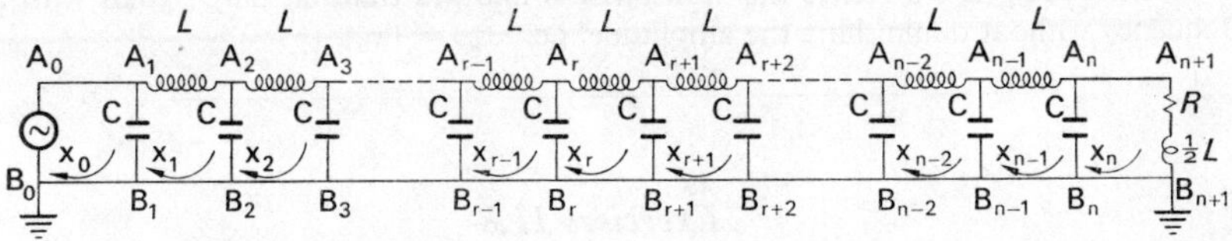

FIG. 11.15

We represent the currents flowing in this network by currents $x_0, x_1, x_2, \ldots, x_r, \ldots, x_{n-1}, x_n$ circulating in the closed loops as indicated in Fig. 11.15. By this means Kirchhoff's first law is satisfied. We apply Kirchhoff's second law to the various loops and obtain

$$\frac{1}{i\omega C}(x_{r+1} - x_r) + i\omega L x_{r+1} + \frac{1}{i\omega C}(x_{r+1} - x_{r+2}) = 0$$

for $r = 0, 1, 2, \ldots, n-2$.

$$\therefore \; x_{r+2} - 2(1 - \tfrac{1}{2} LC\omega^2)\, x_{r+1} + x_r = 0. \tag{1}$$

If $1 - \frac{1}{2}LC\omega^2 > -1$, i.e. $LC\omega^2 < 4$, the difference equation (1) can be written

$$x_{r+2} - 2 \cos \theta x_{r+1} + x_r = 0, \tag{2}$$

where θ is a real angle such that $\cos\theta = 1-\frac{1}{2}LC\omega^2$, $0 \leqslant \theta \leqslant \pi$. The solution of (2) is

$$x_r = A\,e^{ir\theta}+B\,e^{-ir\theta}. \tag{3}$$

By putting $r = 0$ we obtain

$$x_0 = A+B. \tag{4}$$

We apply Kirchhoff's second law to the loop $B_nA_nA_{n+1}B_{n+1}$ and find

$$Rx_n+\frac{1}{2}\,i\omega Lx_n+\frac{1}{i\omega C}\,(x_n-x_{n-1}) = 0.$$

Since $RC\omega = \pm\sin\theta$,

$$x_{n-1} = x_n(1-\tfrac{1}{2}LC\omega^2+iRC\omega) = x_n(\cos\theta\pm i\sin\theta).$$

i.e.

$$x_{n-1} = x_n\,e^{\pm i\theta}. \tag{5}$$

Substituting from (3) into (5), we obtain

$$A\,e^{in\theta}+B\,e^{-in\theta} = e^{\mp i\theta}\{A\,e^{i(n-1)\theta}+B\,e^{-i(n-1)\theta}\}.$$

Using the upper sign we find, with the help of (3),

$$A\,e^{in\theta}+B\,e^{-in\theta} = A\,e^{i(n-2)\theta}+B\,e^{-in\theta}; \quad A = 0, \quad B = x_0,$$

Using the lower sign we find

$$A\,e^{in\theta}+B\,e^{-in\theta} = A\,e^{in\theta}+B\,e^{-i(n-2)\theta}; \quad B = 0, \quad A = x_0.$$

$$\therefore\ x_n = x_0\,e^{\mp in\theta}. \tag{6}$$

The conditions $\cos\theta = 1-\frac{1}{2}LC\omega^2$, $RC\omega = \pm\sin\theta$ imply that either $\omega = 0$ or $\omega^2 = 4(L-CR^2)/(L^2C)$. Hence this transmission line will transmit only signals with this frequency without diminishing the amplitude, i.e. $|x_n| = |x_0|$.

Exercises 11.8

1. A uniform cable has constant resistance R, capacity C and inductance L per unit length. Prove that the potential ϕ at a point at distance x along the cable from a fixed point satisfies the equation

$$\frac{\partial^2\phi}{\partial x^2} = LC\,\frac{\partial^2\phi}{\partial t^2}+RC\,\frac{\partial\phi}{\partial t}\,.$$

An alternating e.m.f. of amount $E\cos pt$ is applied at the end $x = 0$, and the other end $x = l$ is connected to earth through an instrument of complex impedance $k(R+ipL)$. Show that the phase difference between the potentials at the two ends of the cable is

$$\arg\left\{\frac{kn}{kn\cosh nl+\sinh nl}\right\},$$

where $n^2 = -p^2LC+ipRC$.

2. A potential $E\sin\omega t$ is applied at $x = 0$ to a cable that stretches from $x = 0$ to $x = \infty$ and has an earth return. Show that it induces a periodic potential at distance x, whose value at time t is

$$E\,e^{-\alpha x} \sin(\omega t - \beta x),$$

where $$(\alpha + i\beta)^2 = (R + i\omega L)(D + i\omega C).$$

3. A transmission line of length l, for which $R/L = G/C$, is initially at unit potential. At $t = 0$ the end $x = 0$ is earthed. Show that at subsequent time t the potential at x is

$$\frac{4}{\pi} e^{-\varrho t} \sum_{n=0}^{\infty} \frac{1}{2n+1} \sin \frac{(2n+1)\pi x}{2l} \cos \frac{(2n+1)\pi \alpha t}{2l},$$

where $\varrho = R/L$ and $\alpha = (LC)^{-1/2}$.

Miscellaneous Exercises XI

1. State Maxwell's equations for an isotropic conducting medium of conductivity σ, permeability μ, and permittivity ε and show that the interior of such a conductor may be assumed to be uncharged.

Show also that the propagation of electromagnetic waves, of period $2\pi/\omega$, in such a medium is the same as for propagation in a transparent medium of complex permittivity ε' given by

$$\varepsilon' = \varepsilon - \sigma i/\omega.$$

The field vectors of a train of plane polarized electromagnetic waves, travelling in the above medium, at a field point of position vector $\boldsymbol{r}$ are given by

$$\boldsymbol{E} = \boldsymbol{A}_0 \exp\{i\omega(t - \boldsymbol{u}\cdot\boldsymbol{r})\}$$
$$\boldsymbol{H} = \boldsymbol{B} \exp\{i\omega(t - \boldsymbol{u}\cdot\boldsymbol{r})\}$$

where $\boldsymbol{B} = \boldsymbol{B}_0 e^{i\beta}$, $\boldsymbol{u} = \boldsymbol{u}_0 e^{i\lambda}$ and the resolutes of $\boldsymbol{A}_0$, $\boldsymbol{B}_0$ and $\boldsymbol{u}_0$ are all real. Show that $\boldsymbol{u}$, $\boldsymbol{E}$ and $\boldsymbol{H}$ form a right-handed triad. Show also that

$$\beta = \lambda = -\tfrac{1}{2} \tan^{-1}(\sigma/\varepsilon\omega),$$
$$|\boldsymbol{u}_0|^{-1} = c(\varepsilon_0\mu)^{-1/2} \quad \text{where} \quad \varepsilon_0^2 = \varepsilon^2 + \sigma^2/\omega^2.$$

2. An infinite conductor fills the region $z < 0$ and has a plane face $z = 0$. The region between the planes $z = 0$ and $z = a$ is filled with uniform isotropic dielectric material having dielectric constant K. A plane electromagnetic wave, whose wavelength is $(2\pi/k)$ propagates freely in the region $z > a$ so that it is normally incident on the surface of the dielectric. Prove that the reflected wave in the region $z > a$ is either in phase with the incident wave, or has the opposite phase to the incident wave, provided

$$\sqrt{K} \tan(ka + \tfrac{1}{2}p\pi) = \tan(ka\sqrt{K}),$$

where $p = 0$, or 1.

3. Obtain from Maxwell's equations for a homogeneous medium of dielectric constant K, conductivity σ and magnetic permeability μ the "equation of telegraphy" satisfied by the electric vector $\boldsymbol{E}$, and obtain a similar equation for the magnetic vector $\boldsymbol{H}$.
Discuss the propagation in this medium of plane polarized waves of period $2\pi/\omega$, and determine the wave velocity and absorption coefficient.
If the medium is highly conducting, show that the $\boldsymbol{E}$ waves and $\boldsymbol{H}$ waves are out of phase by approximately $\pi/4$, and that the "depth of penetration" is $\sqrt{(\lambda c/\mu\sigma)}$, where λ is the wavelength in free space.

4. Prove that, if an electromagnetic field independent of y exists in the space $0 \leqslant z \leqslant a$ between perfectly conducting planes, then

$$\boldsymbol{E} = \frac{\partial \psi}{\partial z}\boldsymbol{i} - \frac{\partial \psi}{\partial x}\boldsymbol{k}, \quad \boldsymbol{H} = -\frac{1}{c}\frac{\partial \psi}{\partial t}\boldsymbol{j}$$

satisfy Maxwell's equations, provided that $\psi(x, z, t)$ satisfies the wave equation.

If $\psi = \psi_0(z)\, e^{i(\beta x - \nu t)}$, where ψ_0 is chosen to satisfy the boundary condition (tangential electric force zero over $z = 0$ and $z = a$), show that, if $\nu > \beta c$,

$$\nu^2/c^2 - \beta^2 = \omega^2 = r^2\pi^2/a^2, \quad (r = 1, 2, 3, \ldots),$$

and that any one of these solutions corresponds to the portions $0 \leqslant z \leqslant a$ of two unlimited trains of plane waves propagated with speed c in the directions of unit vectors

$$\frac{c}{\nu}\{\omega \boldsymbol{i} \pm \beta \boldsymbol{k}\}.$$

5. Show that the field defined by

$$\boldsymbol{E} = -\frac{\partial \boldsymbol{A}}{\partial t}, \quad \boldsymbol{B} = \mathbf{curl}\, \boldsymbol{A}, \quad \boldsymbol{A} = \{A_0 \sin \alpha y \cos(\omega t - \gamma z)\ 0\ 0\},$$

satisfies Maxwell's equations for free space provided $\alpha^2 + \gamma^2 = \omega^2/c^2$.

Establish conditions on α and ω which must be satisfied so that these expressions can represent a possible field propagating without attenuation in the region between two perfect conductors with surfaces $y = 0$ and $y = a$.

6. A surface current of density $\boldsymbol{i}$ flows in a metal sheet occupying the plane $z = 0$ and the rest of space is empty.

By integrating Maxwell's equation for **curl** $\boldsymbol{H}$ over a suitable surface, or otherwise, express $\boldsymbol{i}$ in terms of the discontinuity in the magnetic field at $z = 0$. State the other conditions which must hold.

Verify that a possible solution of Maxwell's equations is, using cartesian coordinates,

$$\boldsymbol{E} = \{0\ 0\ A \sin n(x - ct)\}$$
$$\boldsymbol{H} = \{0\ -\lambda A \sin n(x - ct)\ 0\}$$

for $z > 0$, where $\lambda = 1/(\mu_0 c)$, and $\boldsymbol{E} = \boldsymbol{H} = \boldsymbol{0}$ for $z < 0$. Determine the current in the sheet necessary to maintain the oscillation.

7. A general dielectric medium is divided into two regions, denoted 1 and 2, by a metal sheet S which carries a current of surface density $\boldsymbol{i}$. By integrating Maxwell's equation for **curl** $\boldsymbol{H}$ over a suitable infinitesimal surface, or otherwise, show that $\boldsymbol{i}$ is related to the discontinuity $(\boldsymbol{H}_{t_2} - \boldsymbol{H}_{t_1})$ in the tangential component of $\boldsymbol{H}$ across S by the formula

$$\boldsymbol{H}_{t_2} - \boldsymbol{H}_{t_1} = \boldsymbol{i} \times \hat{\boldsymbol{n}},$$

where $\hat{\boldsymbol{n}}$ is a unit vector along the normal to S directed from region 1 into region 2.

The sheet S is in the shape of an infinite cylinder of radius a, whose axis lies along the z-axis; the space inside and outside S is empty. Verify that a possible solution of Maxwell's equations is, using cylindrical polar coordinates (r, θ, z),

$$\boldsymbol{H} = \frac{cA}{r}\cos w(t - z/c)\, \hat{\boldsymbol{\theta}},$$

$$\boldsymbol{E} = \frac{A}{\varepsilon_0 r}\cos w(t - z/c)\, \hat{\boldsymbol{r}}$$

for $r > a$, and $\boldsymbol{H} = \boldsymbol{0}, \boldsymbol{E} = \boldsymbol{0}$ for $r < a$, where $\hat{\boldsymbol{r}}$ and $\hat{\boldsymbol{\theta}}$ are unit vectors in the directions of increasing r and θ. Determine the current in the sheet necessary to maintain this oscillation.

$$\left[\text{div}\,\boldsymbol{F} = \frac{1}{r}\frac{\partial}{\partial r}(rF_r) + \frac{1}{r}\frac{\partial F_\theta}{\partial \theta} + \frac{\partial F_z}{\partial z},\right.$$

$$\left.\textbf{curl}\,\boldsymbol{F} = \left(\frac{1}{r}\frac{\partial F_z}{\partial \theta} - \frac{\partial F_\theta}{\partial z}\right)\hat{\boldsymbol{r}} + \left(\frac{\partial F_r}{\partial z} - \frac{\partial F_z}{\partial r}\right)\hat{\boldsymbol{\theta}} + \left(\frac{1}{r}\frac{\partial (rF_\theta)}{\partial r} - \frac{1}{r}\frac{\partial F_r}{\partial \theta}\right)\hat{\boldsymbol{z}}.\right]$$

8. Prove that in an electromagnetic field independent of the coordinate y, where the axes Ox, Oy, Oz are rectangular cartesian, and $\boldsymbol{i}, \boldsymbol{j}, \boldsymbol{k}$ are unit vectors along them, the vectors

$$\boldsymbol{E} = \frac{\partial S}{\partial z}\boldsymbol{i} - \frac{\partial S}{\partial x}\boldsymbol{k}, \qquad \boldsymbol{H} = -\beta\frac{\partial S}{\partial t}\boldsymbol{j}$$

satisfy Maxwell's equations provided that S satisfies the wave equation

$$\frac{\partial^2 S}{\partial x^2} + \frac{\partial^2 S}{\partial z^2} = \frac{1}{c^2}\frac{\partial^2 S}{\partial t^2},$$

where β is a constant, to be found, $\partial S/\partial z = 0$, and c is the velocity of light.

Find an electromagnetic field in the region of space $0 \leqslant x \leqslant a$, $0 \leqslant z \leqslant b$, so chosen that the tangential electric field vanishes over the planes $x = 0$, $x = a$, $z = 0$, $z = b$, and show that, if the frequency is $\omega/2\pi$, then

$$\frac{\omega^2}{c^2} = \frac{r^2\pi^2}{a^2} + \frac{s^2\pi^2}{b^2}, \qquad (r, s = 1, 2, 3, \ldots).$$

9. Show that Maxwell's equations for an isotropic homogeneous non-conducting charge-free medium can be satisfied by taking

$$\boldsymbol{E} = \text{Re}\left(-\frac{\partial}{\partial t}\nabla\times\phi\boldsymbol{a}\right), \qquad \boldsymbol{B} = \text{Re}\,(\nabla\times\nabla\times\phi\boldsymbol{a})$$

where $\boldsymbol{a}$ is a constant unit vector and ϕ satisfies the wave equation.

Taking $\boldsymbol{a}$ in the z-direction, show that the wave equation in cylindrical polar coordinates (R, θ, z) has an axially symmetric solution

$$\phi = J_0(\varkappa R)e^{i(kz-\omega t)}$$

where $\varkappa^2 = \omega^2/c^2 - k^2$ and J_0 denotes the Bessel function of zero order.

For a hollow waveguide with a perfectly conducting cylindrical boundary $R = a$, show that the solution represents *transverse electric* wave modes. Find the permissible values of k for a given frequency and show that the critical frequency ω_c for a given k satisfies the relation $\omega_c^2 = \omega^2 - c^2k^2$.

10. For a certain electromagnetic field in a non-conducting dielectric the scalar potential is zero and the vector potential is $A_r = 0$, $A_\phi = 0$, $A_z = f(r)\,g(\phi)e^{i\alpha(z-ut)}$ in cylindrical coordinates r, ϕ, z. Determine the electric and magnetic field intensities.

Show that electromagnetic waves of this type can be propagated along a dielectric cylinder with perfectly conducting boundaries at $r = a$, $\phi = 0$ and $\phi = \pi/2$ with a velocity $u\{1+(S^2/\alpha^2)\}^{1/2}$, where u is the velocity of light in the dielectric and S is a root of $J_{2m}(Sa) = 0$, where m is a positive integer.

11. Show that Maxwell's equations for a vacuum possess a solution of the form

$$E_x = \partial u/\partial y, \quad E_y = -\partial u/\partial x, \quad E_z = 0,$$
$$H_x = 0, \quad H_y = 0, \quad H_z = \varepsilon_0\, \partial u/\partial t,$$

where u satisfies the equation

$$\frac{\partial^2 u}{\partial x^2}+\frac{\partial^2 u}{\partial y^2} = \frac{1}{c^2}\frac{\partial^2 u}{\partial t^2}.$$

Obtain an integral of this equation in the form

$$u = e^{-ikct}f(x, y) = e^{ik(y-ct)} \int_0^{\sqrt{(r+y)}} e^{-iks^2}\, ds,$$

where k is a constant and $r^2 = x^2+y^2$.

Prove that the wave function

$$e^{-ikct}\cos ky + e^{-ikct}(k/\pi)^{1/2}e^{i\pi/4}[f(x, y)+f(x, -y)]$$

furnishes a solution of the problem of the diffraction of an infinite, plane, monochromatic beam of light, incident normally on the semi-infinite perfectly reflecting plane, $y = 0$, $x \geqslant 0$.

12. In an electromagnetic field which is independent of z new coordinates ξ, η ($\xi \geqslant 0$) are introduced where $x = \frac{1}{2}(\xi^2-\eta^2)$, $y = \xi\eta$. Find those electric fields which are time-harmonic with frequency $\omega/2\pi$ and which have the form $E_x = 0$, $E_y = 0$, $E_z = f(\xi)e^{-ik\eta^2/2}$, where $k = \omega/c$.

The plane wave $E_x = 0$, $E_y = 0$, $E_z = e^{ikx}$ is incident on the perfectly conducting parabolic cylinder $y^2+2x = 1$ from $x > 0$. Find the reflected wave.

CHAPTER 12

THE LORENTZ INVARIANCE OF MAXWELL'S EQUATIONS

12.1 Groups of transformations

We have already considered how some of the electromagnetic quantities transform when we make the coordinate transformations

$$t' = \beta(t - vx/c^2), \quad x' = \beta(x - vt), \quad y' = y, \quad z' = z,$$

where

$$\beta = (1 - v^2/c^2)^{-1/2},$$

and we write the equations in t first (as we shall for the rest of this chapter). It is now time to take up the general question of invariance. A scientific theory is, generally speaking, invariant under some *group* of transformations. The reader may be reminded that by a group is meant a set of quantities between which a binary operation is defined. (This binary operation, often regarded as a product, for transformations consists of applying two transformations in succession.) The binary operation is associative and for the existence of a group there must be an identity element and a reciprocal of every element. In the case of transformations the identity element is the identity transformation and the reciprocal element is the inverse transformation. These ideas are already familiar in the case of the orthogonal group in three dimensions (i.e. transformations from one set of orthogonal axes to another), under which Euclidean geometry is invariant. Indeed the whole invariance under this group is automatically built into a theory as soon as it is expressed in vectorial form. Accordingly the theory of electromagnetism must be invariant under some group which has the orthogonal group in three dimensions as a subgroup. Exactly what group this is we shall return to shortly. Before that, let us consider exactly what it means to say that the theory is *invariant*.

We can envisage using different coordinate systems, connected by a group of transformations, and corresponding to each coordinate system the physical situation is represented by a set of numbers. For example, in the case of electromagnetic theory the numbers might be the six components of the electric field and the magnetic flux density, ***E*** and ***B***. Imagine then three coordinate systems

$$S_1, \quad S_2, \quad S_3$$

with the corresponding sets of numbers

$$\phi_1, \quad \phi_2, \quad \phi_3 .$$

Here ϕ_1 denotes a *set* of numbers—for instance, it might be the six components of ***E*** and ***B*** in the coordinate system S_1. The coordinate systems are related by certain transformations which can be represented diagrammatically by

$$S_1 \xrightarrow{T_{12}} S_2 \xrightarrow{T_{23}} S_3 .$$

As a result the sets of numbers are transformed by corresponding transformations

$$\phi_1 \xrightarrow{t} \phi_2 \xrightarrow{u} \phi_3 .$$

Now the transformations of coordinates obviously have as a result a single transformation

$$S_1 \xrightarrow{T_{13}} S_3 ,$$

and this can be thought of as giving rise to a single transformation of the numbers

$$\phi_1 \xrightarrow{v} \phi_3 .$$

Of course from the mere definition of the transformation of coordinates it follows that

$$T_{23}T_{12} = T_{13}$$

where the product is meant to be read from right to left in accordance with the usual method of writing transformations, i.e. we write

$$S_2 = T_{12}S_1 ,$$
$$S_3 = T_{23}S_2$$

so that

$$S_3 = T_{23}(T_{12}S_1) = (T_{23}T_{12})S_1 .$$

In such circumstances it is to be expected that, if and only if the numbers really represent properties of the physical system, and not merely properties of the coordinate system, then the corresponding transformations for them will satisfy $ut = v$. On the other hand, if they are merely numbers representing in part the properties of the coordinate system, it must make a difference whether one proceeds from the first to the third system directly or via some intermediate stage, and in this case the conditions $ut = v$ will not be fulfilled. The condition just expressed on the transformations of the sets of numbers is known technically as requiring the transformations to be a *representation* of the original group and the argument which we have just given (which is the general form of the *principle of relativity*) can be put by saying that any physically significant set of numbers must transform under a representation of the group of transformations of the theory.

The next problem is how to find such representations. It is instructive to look first at the example of the orthogonal group in three dimensions. Here we know of one example of a representation in the components of an ordinary vector. We may write

$$\boldsymbol{u} = \sum_{i=1}^{3} u_i \boldsymbol{e}_i$$

where $\boldsymbol{e}_1, \boldsymbol{e}_2, \boldsymbol{e}_3$ are taken as unit vectors along the three axes, and when we make a coordinate transformation, say by rotating the unit vectors along the three axes to three new directions $\boldsymbol{e}_i'$ we have

$$\boldsymbol{u} = \sum_{i=1}^{3} u_i' \boldsymbol{e}_i' .$$

Such a rotation must be expressible, however, in the form

$$\boldsymbol{e}_i' = \sum_{j=1}^{3} l_{ij} \boldsymbol{e}_j ,$$

since the new unit vectors have certain components along the old ones, and this is all that is expressed by this equation. However, not all linear transformations of this kind are permitted, since we have to ensure that the new set of unit vectors are again at right angles and the condition for this is easily derived as follows: both the original and the new unit vectors satisfy

$$\boldsymbol{e}_i \cdot \boldsymbol{e}_j = \delta_{ij}, \quad \boldsymbol{e}_i' \cdot \boldsymbol{e}_j' = \delta_{ij}$$

where $\quad \delta_{ij} = 0 \quad \text{if} \quad i \neq j, \quad \text{and} \quad \delta_{ij} = 1 \quad \text{if} \quad i = j.$

Hence

$$\sum_{p,q} l_{ip} l_{jq} \boldsymbol{e}_p \cdot \boldsymbol{e}_q = \sum_{p} l_{ip} l_{jp} = \delta_{ij} .$$

This is the condition for the linear transformation to be an orthogonal one. Returning now to the original vectors it is clear that the transformation of components is determined by

$$\sum_j u_j \boldsymbol{e}_j = \sum_{i,j} (u'_i l_{ij}) \boldsymbol{e}_j,$$

i.e.

$$u_j = \sum_i l_{ij} u'_i,$$

which, by using the condition of the transformation, can be rewritten in the form

$$\sum_j l_{pj} u_j = \sum_{i,j} l_{pj} l_{ij} u'_i = \sum_i \delta_{pi} u'_i = u'_p$$

(cf. the equation for $\boldsymbol{e}'_i$ above). In this case, then, we see that the components of the vectors transform in the same way as the unit vectors themselves, but this is a coincidence.

What has basically been done here is to derive a representation of the group by regarding a vector as a displacement; in other words, one chooses something which, from its geometrical or physical interpretation, is known to be independent of the coordinate system and so to transform under a representation of the group, and uses its representation to define a whole class of such objects. In this way one could construct a series of more complicated representations. For example, continuing with the orthogonal group in three dimensions one could consider the array of quantities

$$A_{ij} = u_i v_j$$

where u_i, v_j are both components of vectors, and then evidently this array will transform under the rule

$$A'_{ij} = \sum_{p,\, q=1}^{3} l_{ip} l_{jq} A_{pq},$$

which must from its construction define a representation of the group, as the reader may verify. These more complicated transformations are called the tensor representations of the group. The particular transformation above is of a tensor of rank 2. By taking products of more than two vectors we get the higher order tensor representations.

These are by no means all the possible representations, although they are important ones, as can be seen from the following example. Consider, for simplicity, a tensor of rank 2 in *two dimensions* for which the orthogonal transformations can be represented by [(for a vector u_1, u_2)]:

$$\begin{aligned} u'_1 &= u_1 \cos\theta + u_2 \sin\theta, \\ u'_2 &= -u_1 \sin\theta + u_2 \cos\theta. \end{aligned} \tag{12.1}$$

Writing out the transformation of the tensor at length it is easy to derive

$$\begin{aligned}
A'_{11} &= A_{11}\cos^2\theta+(A_{21}+A_{12})\cos\theta\sin\theta+A_{22}\sin^2\theta,\\
A'_{12} &= A_{12}\cos^2\theta-A_{21}\sin^2\theta+(A_{22}-A_{11})\cos\theta\sin\theta,\\
A'_{21} &= A_{21}\cos^2\theta-A_{12}\sin^2\theta+(A_{22}-A_{11})\cos\theta\sin\theta,\\
A'_{22} &= A_{11}\sin^2\theta-(A_{21}+A_{12})\cos\theta\sin\theta+A_{22}\cos^2\theta,
\end{aligned}$$

from which it follows that

$$A'_{11}+A'_{22} = A_{11}+A_{22}\,.$$

It is already well known, and is again obvious from the formulae, that another invariant is given by $A_{12}-A_{21}$, so that the parts of the tensor which actually transform are only two in number, and their transformation can be conveniently represented by

$$\begin{aligned}
A'_{12}+A'_{21} &= (A_{12}+A_{21})\cos 2\theta+(A_{22}-A_{11})\sin 2\theta,\\
A'_{22}-A'_{11} &= -(A_{21}+A_{12})\sin 2\theta+(A_{22}-A_{11})\cos 2\theta.
\end{aligned} \tag{12.2}$$

It is a very striking fact that this transformation is of exactly the same form as that originally assumed for the vectors except that the angle of rotation is replaced by double its original value. This suggests reversing the whole argument and beginning by considering two quantities ϕ_1, ϕ_1 whose transformation is

$$\begin{aligned}
\phi'_1 &= \phi_1\cos(\theta/2)+\phi_2\sin(\theta/2),\\
\phi'_2 &= -\phi_1\sin(\theta/2)+\phi_2\cos(\theta/2).
\end{aligned} \tag{12.3}$$

By interpreting these as one of the vectors *in the above argument* and introducing ψ_1, ψ_2 for the other, and then halving all the angles, it is then clear that the two quantities $(\phi_1\psi_2+\phi_2\psi_1,\ \phi_2\psi_2-\phi_1\psi_1)$ are the components of a vector *in the two-dimensional space*. The quantities ϕ_1, ϕ_2 are then transformed under a two-valued or spin representation of the group since, when the new vector is rotated through an angle 2π and so returns to its original value, these quantities are rotated through π and so are changed in sign.

12.2 Four-vectors and six-vectors

It is now time to consider the different transformations of sets of numbers corresponding to coordinate systems in uniform relative motion. In this case we have a rather obvious choice to begin with, when we seek for sets of quantities obviously transforming under a representation of the group, for we can choose $(t, \boldsymbol{r})$ as a prototype.

Any set of quantities

$$(\phi, \boldsymbol{A}) = (\phi, A_1, A_2, A_3) \tag{12.4}$$

transforming in the same way will be called a four-vector. We have then, for any four-vectors, the transformation

$$\phi' = \beta(\phi - vA_1/c^2), \quad A_1' = \beta(A_1 - v\phi), \quad A_2' = A_2, \quad A_3' = A_3, \tag{12.5}$$

and we may notice about such a four-vector the important fact that

$$c^2\phi^2 - \boldsymbol{A}^2$$

is also unchanged by the transformation. In the particular case of the time and space coordinates the corresponding quantity which is also unchanged is $c^2t^2 - \boldsymbol{r}^2 = c^2t^2 - x^2 - y^2 - z^2$. Corresponding to two four-vectors

$$(\phi, \boldsymbol{A}), \quad (\psi, \boldsymbol{B})$$

we can define their sum as $(\phi + \psi, \boldsymbol{A} + \boldsymbol{B})$, and in view of the linear character of the transformations concerned, this sum will also be a four-vector. It will have a corresponding invariant

$$c^2(\phi + \psi)^2 - (\boldsymbol{A} + \boldsymbol{B})^2$$

and, if we subtract from this the parts $c^2\phi^2 - \boldsymbol{A}^2$, $c^2\psi^2 - \boldsymbol{B}^2$ already found as being invariant, we are left with an important invariant bilinear product of two four-vectors, $c^2\phi\psi - \boldsymbol{A} \cdot \boldsymbol{B}$, which is obviously related to the scalar product in ordinary vector analysis.

The next step is to inquire about analogues of the vector products in ordinary vector analysis. It is, of course, obvious that under the orthogonal group, for example, we could consider, instead of the vector product, the array of nine quantities transforming under the second-order tensor representation but it is not convenient to do so here because electromagnetic theory is not concerned, in general, with such quantities. (They enter, for instance, in rigid mechanics, in the definition of moments and products of inertia, and in certain other branches of applied mathematics, such as elasticity.) Accordingly we shall consider products of $(\phi, \boldsymbol{A})$ and $(\psi, \boldsymbol{B})$ which are made up of the following scalar and vector expressions:

$$\phi\psi, \quad \psi\boldsymbol{A}, \quad \phi\boldsymbol{B}, \quad \boldsymbol{A} \cdot \boldsymbol{B}, \quad \boldsymbol{A} \times \boldsymbol{B}.$$

By the very way in which these are written down they must be automatically invariant under the orthogonal group in three dimensions but we are concerned also in transforming between coordinate systems in uniform relative motion. Writing down the transformations we get

$$\psi' A_1' = \beta^2[\psi A_1 - vA_1B_1/c^2 - v\psi\phi + v^2\phi B_1/c^2], \tag{12.6}$$
$$\psi' A_2' = \beta[\psi A_2 - vB_1A_2/c^2], \tag{12.7}$$
$$\psi' A_3' = \beta[\psi A_3 - vB_1A_3/c^2], \tag{12.8}$$

and it is at once obvious that, in eqn. (12.7), quantities (B_1A_2) are entering which do not arise in the original set of scalar and vector expressions; similarly B_1A_3 enter eqn. (12.8). Accordingly, the quantity $\psi\boldsymbol{A}$ by itself cannot be part of our product, but by observing the details of its transformation it suggests that we should consider instead $\phi\boldsymbol{B} - \psi\boldsymbol{A}$. When we do this we get

$$\phi' B_1' - \psi' A_1' = \phi B_1 - \psi A_1, \tag{12.9}$$
$$\phi' B_2' - \psi' A_2' = \beta(\phi B_2 - \psi A_2) - \frac{\beta v}{c^2}(A_1B_2 - A_2B_1), \tag{12.10}$$
$$\phi' B_3' - \psi' A_3' = \beta(\phi B_3 - \psi A_3) - \frac{\beta v}{c^2}(A_1B_3 - A_3B_1), \tag{12.11}$$

and we now see that we are half-way to a satisfactory formulation of the product. The only remaining difficulty is that the transformation involves, as well as the particular combination of vectors with which we started, the components of the vector product of the two vectors. But the transformation of these,

$$A_2'B_3' - A_3'B_2' = A_2B_3 - A_3B_2, \tag{12.12}$$
$$A_3'B_1' - A_1'B_3' = \beta(A_3B_1 - A_1B_3) + \beta v(\phi B_3 - \psi A_3), \tag{12.13}$$
$$A_1'B_2' - A_2'B_1' = \beta(A_1B_2 - A_2B_1) - \beta v(\phi B_2 - \psi A_2), \tag{12.14}$$

does not introduce any new quantities and so we have succeeded in defining something which does transform under a representation of the group.

If we write the vectors $\boldsymbol{P}, \boldsymbol{Q}$ for

$$\boldsymbol{P} = \phi\boldsymbol{B} - \psi\boldsymbol{A}, \quad \boldsymbol{Q} = \boldsymbol{A} \times \boldsymbol{B},$$

their components in the new frame are given by eqns. (12.9–11) and eqns. (12.12–14) and may be written

$$\begin{aligned} P_1' &= P_1, & Q_1' &= Q_1, \\ P_2' &= \beta\left(P_2 - \frac{v}{c^2}Q_3\right), & Q_2' &= \beta(Q_2 + vP_3), \\ P_3' &= \beta\left(P_3 + \frac{v}{c^2}Q_2\right), & Q_3' &= \beta(Q_3 - vP_2). \end{aligned} \tag{12.15}$$

We can now call any quantity $(\boldsymbol{P}, \boldsymbol{Q})$ transforming like $(\phi\boldsymbol{B}-\psi\boldsymbol{A}, \boldsymbol{A}\times\boldsymbol{B})$ ***a six-vector***.

It is interesting to notice that for transformations in which the velocity v is small compared with that of light, eqns. (12.15) reduce to

$$\boldsymbol{P}' \approx \boldsymbol{P}, \quad \boldsymbol{Q}' = \boldsymbol{Q}-\boldsymbol{V}\times\boldsymbol{P}.$$

Example 1. Prove that $(\operatorname{div}\boldsymbol{P}, \mathbf{curl}\,\boldsymbol{Q}-\partial\boldsymbol{P}/\partial t)$ is a four-vector if $(\boldsymbol{P}, \boldsymbol{Q})$ is a six-vector. Prove also that $\boldsymbol{P}^2-\boldsymbol{Q}^2/c^2$, $\boldsymbol{P}\cdot\boldsymbol{Q}$ are invariant.

An immediate calculation gives

$$\begin{aligned}\operatorname{div}'\boldsymbol{P}' &= \left(\frac{\partial x}{\partial x'}\frac{\partial}{\partial x}+\frac{\partial t}{\partial x'}\frac{\partial}{\partial t}\right)P_1+\beta\frac{\partial}{\partial y}\left(P_2-\frac{v}{c^2}Q_3\right)+\beta\frac{\partial}{\partial z}\left(P_3+\frac{v}{c^2}Q_2\right)\\ &= \beta\left[\operatorname{div}\boldsymbol{P}+\frac{\beta v}{c^2}\frac{\partial P_1}{\partial t}-\frac{\beta v}{c^2}\left(\frac{\partial Q_3}{\partial y}-\frac{\partial Q_2}{\partial z}\right)\right]\\ &= \beta\left[\operatorname{div}\boldsymbol{P}-\frac{v}{c^2}\left(\mathbf{curl}\,\boldsymbol{Q}-\frac{\partial\boldsymbol{P}}{\partial t}\right)_1\right],\end{aligned}$$

which is the transformation for the "time-component" of a four-vector. The remaining results follow directly from the transformations.

Example 2. If $\boldsymbol{P}, \boldsymbol{Q}$ is a six-vector, and $(\phi, \boldsymbol{A})$ is a four-vector, prove that

$$(\boldsymbol{A}\cdot\boldsymbol{P}, \ \boldsymbol{A}\times\boldsymbol{Q}+c^2\phi\boldsymbol{P})$$

is also a four-vector.

First consider $\boldsymbol{A}\cdot\boldsymbol{P}$.

Then

$$\begin{aligned}\boldsymbol{A}'\cdot\boldsymbol{P}' &= A_1'P_1'+A_2'P_2'+A_3'P_3'\\ &= \beta P_1A_1-\beta P_1v\phi+\beta\left(P_2-\frac{v}{c^2}Q_3\right)A_2+\beta\left(P_3+\frac{v}{c^2}Q_2\right)A_3\\ &= \beta\left\{\boldsymbol{P}\cdot\boldsymbol{A}-\frac{v}{c^2}(A_2Q_3-A_3Q_2+c^2\phi P_1)\right\}.\end{aligned}$$

This is the transformation of the time-component of a four-vector if the quantity in brackets, $A_2Q_3-A_3Q_2+c^2\phi P_1$ is the x-component; and in fact it is the x-component of $\boldsymbol{A}\times\boldsymbol{Q}+c^2\phi\boldsymbol{P}$.

The calculations of the transformation of the remaining components are carried out in similar manner.

Example 3. (a) Prove that the operator

$$\left(\frac{1}{c^2}\frac{\partial}{\partial t}, \ -\nabla\right)$$

is a four-vector. Use this to deduce the first part of Example 1 again.

(b) If $(\phi, \boldsymbol{A})$ is a four-vector, deduce that

$$\frac{\partial\phi}{\partial t}+\operatorname{div}\boldsymbol{A}$$

is invariant and that

$$\left\{\frac{1}{c^2}\left(\frac{\partial A}{\partial t}+\nabla[\phi c^2]\right),\ -\nabla\times A\right\}$$

is a six-vector.

(a)
$$\frac{1}{c^2}\frac{\partial}{\partial t'} = \frac{1}{c^2}\left(\frac{\partial t}{\partial t'}\frac{\partial}{\partial t}+\frac{\partial x}{\partial t'}\frac{\partial}{\partial x}\right)$$

$$= \beta\left[\frac{1}{c^2}\frac{\partial}{\partial t}-\frac{v}{c^2}\left(-\frac{\partial}{\partial x}\right)\right],$$

$$\left(-\frac{\partial}{\partial x'}\right) = \beta\left[\left(-\frac{\partial}{\partial x}\right)-v\left(\frac{1}{c^2}\frac{\partial}{\partial t}\right)\right],$$

which proves the first result. Then Example 1 follows, by using the result of Example 2, with (ϕ, A) taken as

$$\left(\frac{1}{c^2}\frac{\partial}{\partial t},\ -\nabla\right).$$

(b) The first expression is obviously the invariant derived from

$$\left(\frac{1}{c^2}\frac{\partial}{\partial t},\ -\nabla\right), \quad \text{and} \quad (\phi, A).$$

The second is the six-vector constructed from them.

12.3 The Lorentz group

We must now inquire exactly what is the group with which we are concerned here. We know already that it has the orthogonal group in three dimensions as a subgroup. In fact we can prove that the set of all rotations, together with all Lorentz transformations of the kind considered before, that is transformations of the form

$$t' = \beta\left(t-\frac{vx}{c^2}\right),\quad x' = \beta(x-vt),\quad y' = y,\quad z' = z \quad \text{with } \beta = \left(1-\frac{v^2}{c^2}\right)^{-1/2},$$

do form a group. In order to establish this we have to show that the product of any two elements of the group again belongs to the group. As far as the product of two rotations is concerned this is well known and amounts to the theorem that any member of the rotation group is a rotation in the elementary sense about a certain axis (Euler's theorem). If we are concerned with the product of a Lorentz transformation and a rotation, it is clear from the way in which we are able to write our Lorentz transformation in vectorial form that this will again lie in the group. Indeed we tacitly take account of this whenever we choose the x-axis as the direction of separation of the two coordinate systems; for this is equivalent to performing a rotation of the axes so that the given direction is along the common x-axis. It only remains therefore to consider the product of two Lorentz transformations. Two cases arise depending on whether the direction of the velocity in each case is the same or different. If it is the same, we have for the first transformation

$$t' = \beta_u(t-ux/c^2), \quad x' = \beta_u(x-ut),$$

and then applying a second transformation we have

$$t'' = \beta_u\beta_v\left(1+\frac{uv}{c^2}\right)\left(t-\frac{u+v}{1+uv/c^2}\,\frac{x}{c^2}\right), \tag{12.16}$$

$$x'' = \beta_v\beta_u\left(x-ut-vt+\frac{uv}{c^2}\,x\right),$$

$$= \beta_u\beta_v\left(1+\frac{uv}{c^2}\right)\left(t-\frac{u+v}{1+uv/c^2}\,t\right). \tag{12.17}$$

In order to establish that the resultant transformation still belongs to the group it is clear from these equations that we only need to show that

$$\beta_V = \beta_u\beta_v\left(1+\frac{uv}{c^2}\right),$$

where

$$\beta_V = (1-V^2/c^2)^{-1/2}, \quad V = \frac{u+v}{1+uv/c^2} \tag{12.18}$$

and corresponds to a relative velocity V between the frames. The reader may instantly verify this by using the definitions.

If the two velocities are not in the same direction, we consider first the case in which they are perpendicular, and we choose these directions as the x- and y-axes. The first transformation becomes

$$t' = \beta_u(t-ux/c^2), \quad x' = \beta_u(x-ut), \quad y' = y.$$

The second transformation as well gives the result

$$t'' = \beta_v\{\beta_u t-(u\beta_u x+vy)/c^2\}, \tag{12.19}$$

$$x'' = \beta_u(x-ut), \tag{12.20}$$

$$y'' = \beta_v(y-v\beta_u t+\beta_u uvx/c^2). \tag{12.21}$$

Since the z-coordinates are unchanged in both of these transformations we need not consider them at all in this piece of work. In order to show that the transformation t, x, y to t'', x'', y'' is a transformation of the group we rearrange the relations, by a suitable choice of $V, \xi, \eta, \xi'', \eta''$ in the form

$$t'' = \beta_V(t - V\xi/c^2), \quad \xi'' = \beta_V(\xi - Vt), \quad \eta'' = \eta. \tag{12.22–24}$$

First we consider the coefficient of t in the equivalent relations (12.22) and (12.19). We must have

$$\beta_V = \beta_u\beta_v, \quad \text{i.e.} \quad 1 - V^2/c^2 = (1 - u^2/c^2)(1 - v^2/c^2).$$

The latter relation can be written in the alternative forms:

$$V^2 = u^2 + v^2(1 - u^2/c^2) = u^2 + v^2/\beta_u^2 \tag{12.25}$$

or

$$V^2 = v^2 + u^2(1 - v^2/c^2) = v^2 + u^2/\beta_v^2. \tag{12.26}$$

The remaining terms, concerned with space coordinates, give

$$\beta_V V\xi = \beta_v(u\beta_u x + vy).$$

Therefore

$$\xi = \frac{ux}{V} + \frac{vy}{V\beta_u} = x\cos\alpha + y\sin\alpha, \tag{12.27}$$

where

$$\cos\alpha = u/V, \quad \sin\alpha = v/(V\beta_u) \quad \text{or} \quad \tan\alpha = v/(u\beta_u) \tag{12.28}$$

in accordance with eqn. (12.25).

We now eliminate t between eqns. (12.20) and (12.21). This leads to

$$\begin{aligned} v\beta_v x'' - uy'' &= v\beta_u\beta_v x - u\beta_v y - \beta_u\beta_v u^2 vx/c^2 \\ &= v\beta_u\beta_v x(1 - u^2/c^2) - u\beta_v y \\ &= \frac{v\beta_v}{\beta_u}x - u\beta_v y \end{aligned}$$

or

$$vx'' - \frac{uy''}{\beta_v} = \frac{vx}{\beta_u} - uy. \tag{12.29}$$

Guided by the form of eqn. (12.27) and the relations (12.25) and (12.26) we divide (12.29) by $-V$ and obtain

$$-\frac{v}{V}x'' + \frac{u}{V\beta_v}y'' = -\frac{v}{V\beta_u}x + \frac{u}{V}y$$

or

$$-x''\sin\alpha'' + y''\cos\alpha'' = -x\sin\alpha + y\cos\alpha, \tag{12.30}$$

which we identify with (12.24) in the form

$$\eta'' = -x''\sin\alpha'' + y''\cos\alpha'' = -x\sin\alpha + y\cos\alpha = \eta, \tag{12.31}$$

where $\cos\alpha'' = u/(V\beta_v)$, $\sin\alpha'' = v/V$, or $\tan\alpha'' = v\beta_v/u$. (12.32)

Corresponding to the first section of (12.31) we should have

$$\begin{aligned}
\xi'' = x''\cos\alpha + y''\sin\alpha &= \frac{ux''}{V\beta_v} + \frac{vy''}{V} \\
&= \frac{u\beta_u}{V\beta_v}(x-ut) + \frac{v\beta_v}{V}\left(y - v\beta_u t + \frac{\beta_u uvx}{c^2}\right) \\
&= \frac{u\beta_u}{V\beta_v}\left(1+\frac{v^2\beta_v^2}{c^2}\right)x + \frac{v\beta_v}{V}y - \left(\frac{u^2\beta_u}{V\beta_v} + \frac{v^2\beta_u\beta_v}{V}\right)t \\
&= \frac{u\beta_u\beta_v}{V}x + \frac{v\beta_v}{V}y - \frac{\beta_u}{V\beta_v}(u^2+v^2\beta_v^2)t \\
&= \beta_u\beta_v\left(\frac{ux}{V} + \frac{vy}{V\beta_u}\right) - \frac{\beta_u}{V\beta_v}\left(\frac{u^2+v^2-u^2v^2/c^2}{1-v^2/c^2}\right)t \\
&= \beta_V(x\cos\alpha + y\sin\alpha) - \frac{\beta_u\beta_v}{V}V^2t^2 \\
&= \beta_V(\xi - Vt),
\end{aligned}$$

which is eqn. (12.23).

The relations between (ξ, η) and (x, y) and between (ξ'', η'') and (x'', y'') are both rotations through the angles α, α'' respectively with $\alpha \neq \alpha''$. Hence these two successive Lorentz transformations are equivalent to a space-rotation through an angle α, followed by a Lorentz transformation for velocity V along the new axis (ξ-axis), followed by a further rotation through $-\alpha''$ to give the final positions of the x'', y'' axes. The chief importance of this result is that by a suitable choice of u, v the angle α, i.e. the direction of the velocity V, can be given any value and so we deduce that a Lorentz transformation in any direction is equivalent to two such transformations along axes at right angles. This result, combined with that for two transformations in the same direction, suffices to prove that we are dealing with a complete group of transformations. When we speak of the invariance of Maxwell's equations we are referring to their transformation properties under this group of transformations.

Example 1. When the two velocities in perpendicular directions have equal magnitudes we have the special results:

$$V^2 = u^2(2-u^2/c^2);$$
$$\cos\alpha = u/V, \quad \sin\alpha = u/v\beta, \quad \tan\alpha = 1/\beta,$$
$$\cos\alpha'' = u/V\beta, \quad \sin\alpha'' = u/V,$$
$$\tan\alpha'' = \beta; \quad \beta = \beta_u = \beta_v = (1-u^2/c^2)^{-1/2}.$$

Hence $\alpha'' = \pi/2-\alpha$.

Example 2. If V, α are chosen, the velocities along perpendicular axes which give the same transformation are given by

$$u = V\cos\alpha, \qquad V^2 = u^2+v^2-u^2v^2/c^2.$$

Therefore

$$v^2\{1-(V^2/c^2)\cos^2\alpha\} = V^2(1-\cos^2\alpha),$$

i.e.

$$v = V\sin\alpha\{1-(V^2/c^2)\cos^2\alpha\}^{-1/2}.$$

12.4 Maxwell's equations

In order to establish these transformation properties we need certain experimental results. The first result is that the charge, being simply the number of electrons present, is unchanged by the transformation. Let us adopt two coordinate systems, in which the primed coordinate is that in which a small element of charge is at rest. We can write for the total charge

$$\mathrm{e} = \varrho_0\,\mathrm{d}x'\,\mathrm{d}y'\,\mathrm{d}z'. \tag{12.33}$$

Now consider the transformation to another coordinate system in which

$$\mathrm{d}t' = \beta\left(\mathrm{d}t-\frac{V\,\mathrm{d}x}{c^2}\right), \quad \mathrm{d}x' = \beta(\mathrm{d}x-V\,\mathrm{d}t), \quad \mathrm{d}y' = \mathrm{d}y, \quad \mathrm{d}z' = \mathrm{d}z.$$

We are concerned with an element of volume in the new (primed) coordinate system, which is $\mathrm{d}x\,\mathrm{d}y\,\mathrm{d}z$ (with $\mathrm{d}t = 0$) unlike in the primed system where it is $\mathrm{d}x'\,\mathrm{d}y'\,\mathrm{d}z'$ (with $\mathrm{d}t' = 0$). That is to say the volume of the element has the value $\mathrm{d}x'\,\mathrm{d}y'\,\mathrm{d}z'/\beta$. If the total charge is not to be altered by the transformation, then it must follow that

$$\mathrm{e} = \varrho\,\mathrm{d}x\,\mathrm{d}y\,\mathrm{d}z = (\varrho/\beta)\,\mathrm{d}x'\,\mathrm{d}y'\,\mathrm{d}z' = \varrho_0\,\mathrm{d}x'\,\mathrm{d}y'\,\mathrm{d}z',$$

and as a consequence

$$\varrho = \beta\varrho_0. \tag{12.34}$$

It is useful now to relate this to the way in which velocity transforms, because a moving charge also constitutes a current and a product of the charge density and the velocity gives the current density. We have already considered the formulae for transformation of velocity:

$$v_x' = \frac{v_x-V}{1-\dfrac{v_xV}{c^2}}, \qquad v_y' = \frac{1}{\beta_V}\frac{v_y}{1-\dfrac{v_xV}{c^2}}, \text{ and so on,}$$

and in the discussions above about the combination of Lorentz transformations in the same direction we proved [eqn. (12.18)] that

$$\beta_V \beta_v = \frac{\beta_{v'}}{1 - \dfrac{V v_x}{c^2}}.$$

It follows immediately as a consequence of these two results that

$$\beta_{v'} v'_x = \beta_V(\beta_v v_x - \beta_v V), \quad \beta_{v'} v'_y = \beta_v v_y.$$

In other words, the expression $(\beta, \beta v)$ behaves just like the time and space coordinates $(t, \boldsymbol{r})$ under the transformation, and is therefore a four-vector. Accordingly if one defines the charge and current four-vector as $(\varrho, \boldsymbol{J}) = (\varrho, \varrho \boldsymbol{v})$ the expression will have as its scalar (time) component the density *in any frame*. Accordingly we take the charge-current four-vector as the basis needed to estimate the transformation properties of the whole set of equations.

In discussing the invariance of the equations it is necessary to take up again the method of derivation which was discussed in the last volume. We will take for granted the three equations

$$\operatorname{div} \boldsymbol{D} = \varrho, \quad \operatorname{div} \boldsymbol{B} = 0, \quad \mathbf{curl}\, \boldsymbol{E} = -\frac{\partial \boldsymbol{B}}{\partial t}.$$

(It is, of course, an assumption that these particular equations are to be left unaltered, and only the remaining equation is to be changed. At this stage the main reason for this assumption is that it is logically consistent to assume these three equations, whereas the remaining one *cannot* hold for non-steady currents. But ultimately the justification is the agreement of the predictions of the whole set of equations with observation.) The remaining equation was derived from the fact that $\mathbf{curl}\, \boldsymbol{H} \neq \boldsymbol{J}$, by observing that the equations of continuity for the charge-current vector gives

$$\operatorname{div} \boldsymbol{J} = -\frac{\partial \varrho}{\partial t} \neq 0.$$

Now in fact our derivation, which followed Maxwell's argument, was incomplete, since all that is shown by these arguments is that the expression

$$\mathbf{curl}\, \boldsymbol{H} - \frac{\partial \boldsymbol{D}}{\partial t} - \boldsymbol{J}$$

has zero divergence. By well-known theorems in vector analysis it follows that

$$\text{curl}\ \boldsymbol{H} = \frac{\partial \boldsymbol{D}}{\partial t} + \boldsymbol{J} + \text{curl}\ \boldsymbol{G},$$

where we can in addition impose the supplementary equation div $\boldsymbol{G} = 0$. Now it follows from Example 1 of p. 518 that $(\varrho, \boldsymbol{J})$ will be a four-vector if $(\boldsymbol{D}, \boldsymbol{H}-\boldsymbol{G})$ is a six-vector. By a similar argument, $(\boldsymbol{B}, -\boldsymbol{E})$ is also a six-vector. The difference between the equations found by this method, and Maxwell's equations as found in Volume 2, is not in the four main equations at all. For these become identical with the previous set by writing $\boldsymbol{H}' = \boldsymbol{H}-\boldsymbol{G}$ for the magnetic field vector. But the difference appears, of course, in the constitutive relations. In vacuum the introduction of the field $\boldsymbol{G}$ may lead either to the relation

(a) $$\boldsymbol{B} = \mu_0 \boldsymbol{H}'$$

or to (b) $$\boldsymbol{B} = \mu_0 \boldsymbol{H}.$$

If we assume the first of these we have at once

$$\text{curl curl}\ \boldsymbol{H}' = -\frac{1}{\mu_0}\nabla^2 \boldsymbol{B} = -\varepsilon_0 \frac{\partial^2 \boldsymbol{B}}{\partial t^2}$$

in free space so that

$$\frac{1}{c^2}\frac{\partial^2 \boldsymbol{B}}{\partial t^2} - \nabla^2 \boldsymbol{B} = 0, \tag{12.35}$$

where $c^2 \varepsilon_0 \mu_0 = 1$. In the same way

$$\frac{1}{c^2}\frac{\partial^2 \boldsymbol{E}}{\partial t^2} - \nabla^2 \boldsymbol{E} = 0. \tag{12.36}$$

On the other hand, if we assume the second we find

$$\text{curl curl}\ \boldsymbol{H} = -\frac{1}{\mu_0}\nabla^2 \boldsymbol{B} = -\varepsilon_0 \frac{\partial^2 \boldsymbol{B}}{\partial t^2} - \nabla^2 \boldsymbol{G}, \tag{12.37}$$

$$\frac{1}{c^2}\frac{\partial^2 \boldsymbol{B}}{\partial t^2} - \nabla^2 \boldsymbol{B} = -\mu_0 \nabla^2 \boldsymbol{G}, \tag{12.38}$$

$$\frac{1}{c^2}\frac{\partial^2 \boldsymbol{E}}{\partial t^2} - \nabla^2 \boldsymbol{E} = -\mu_0 \frac{\partial}{\partial t}\ \text{curl}\ \boldsymbol{G}. \tag{12.39}$$

The first of these relations, eqns. (12.35) and (12.36), predicts the existence of freely travelling electromagnetic waves and is therefore in agreement with

Hertz's experimental results. The second pair, eqns. (12.38) and (12.39), do not have the form of wave equations, because of the expressions on the right-hand side which depend upon the auxiliary field $\boldsymbol{G}$. Accordingly we use Hertz's experiments to decide in favour of the first alternative and so identify the field which arises *including the auxiliary field* as the magnetic field which we now call H. From the fact that $(\boldsymbol{D}, \boldsymbol{H})$ $(\boldsymbol{B}, -\boldsymbol{E})$ are six-vectors we may write down at once the transformation laws of the electric and magnetic quantities and they are as follows:

$$D_1' = D_1, \qquad H_1' = H_1,$$

$$D_2' = \beta\left(D_2 - \frac{V}{c^2}H_3\right), \qquad H_2' = \beta(H_2 + VD_3), \qquad (12.40), (12.41)$$

$$D_3' = \beta\left(D_3 + \frac{V}{c^2}H_2\right), \qquad H_3' = \beta(H_3 - VD_2),$$

$$B_1' = B_1, \qquad E_1' = E_1,$$

$$B_2' = \beta\left(B_2 + \frac{V}{c^2}E_3\right), \qquad E_2' = \beta(E_2 - VB_3), \qquad (12.42), (12.43)$$

$$B_3' = \beta\left(B_3 - \frac{V}{c^2}E_2\right), \qquad E_3' = \beta(E_3 + VB_2).$$

An immediate consequence of these transformations is the following. We can, by the ordinary rotation of the axes, obviously reduce one of the two vectors which form any six-vector $(\boldsymbol{P}, \boldsymbol{Q})$ to the form $\boldsymbol{Q} = \{0\ 0\ Q_3\}$, and in general the axes can be chosen so that the other one has then the form $\boldsymbol{P} = \{0\ P_2\ P_3\}$. Suppose now that a transformation of coordinates of the usual form is made. As a result these two forms acquire the expressions

$$P_1' = 0, \qquad Q_1' = 0,$$

$$P_2' = \beta\left(P_2 - \frac{V}{c^2}Q_3\right), \qquad Q_2' = \beta VP_3, \qquad (12.44), (12.45)$$

$$P_3' = \beta P_3, \qquad Q_3' = \beta(Q_3 - VP_2).$$

Two different cases now arise. Firstly, let it be supposed that $P_3 = 0$, so that the original two vectors were perpendicular. By choosing

$$V = \frac{c^2 P_2}{Q_3} \quad \text{if} \quad \left|\frac{cP_2}{Q_3}\right| < 1,$$

it follows that

$$\boldsymbol{P} = 0, \qquad (12.46)$$

$$\boldsymbol{Q} = \{0\ 0\ Q_3'\} \qquad (12.47)$$

and the magnitude of the only remaining component of the form is

$$Q_3' = \sqrt{(Q_3^2 - c^2P_2^2)}. \tag{12.48}$$

On the other hand, if

$$V = \frac{Q_3}{P_2} \quad \text{and} \quad \left|\frac{Q_3}{cP_2}\right| < 1,$$

the reduction is carried out the other way round to the form

$$\boldsymbol{Q} = 0, \tag{12.49}$$

$$\boldsymbol{P} = \{0\ P_2'\ 0\}, \tag{12.50}$$

with the sole remaining component

$$P_2' = \sqrt{\left(P_2^2 - \frac{1}{c^2}Q_3^2\right)}.$$

Both of these reductions are impossible in the case of $Q_3^2 = c^2P_2^2$, which is the one arising for plane waves, see Chapter 11. Another reduction, which is always possible, even when $P_3 \neq 0$, is to make the two vectors parallel. From the same equations the condition for this is

$$\frac{P_2 - (V/c^2)Q_3}{VP_3} = \frac{P_3}{Q_3 - VP_2},$$

and this easily reduces to

$$\frac{V/c}{1+V^2/c^2} = \frac{P_2Q_3/c}{P^2 - Q^2/c^2} = \frac{|\boldsymbol{P}\times\boldsymbol{Q}|}{P^2 - Q^2/c^2}.$$

By a simple algebraic manipulation this can be written

$$\left(\frac{1-V/c}{1+V/c}\right)^2 = \frac{P_2^2 + P_3^2 - Q_3^2/c^2 - 2P_2Q_3/c}{P_2^2 + P_3^2 - Q_3^2/c^2 + 2P_2Q_3/c} = \alpha^2,$$

and if α is the positive number determined by this equation we have

$$\frac{V}{c} = \frac{1-\alpha}{1+\alpha}. \tag{12.51}$$

Notice that the above reductions can be carried out either with the six-vector $(\boldsymbol{D}, \boldsymbol{H})$ or with $(\boldsymbol{B}, -\boldsymbol{E})$. The reader can easily verify the equivalence of such reductions in the special case of vacuum, where $\boldsymbol{D} = \varepsilon_0\boldsymbol{E}$, $\boldsymbol{B} = \mu_0\boldsymbol{H}$.

Another interesting application of the transformations is to the experiment of Wilson and Wilson mentioned earlier. This experiment concerns a magnetic dielectric (which was made by embedding small steel balls in sealing wax). This dielectric is between the plates of a moving condenser and the plates of this condenser are short-circuited by means of brushes and a wire which passes through a ballistic galvanometer. The condenser moves with a uniform speed in the direction of the x-axis, the x-y plane being chosen parallel to the plates of the condenser. A magnetic field is then applied in the y-direction (see Fig. 12.1). In the laboratory coordinate system it follows

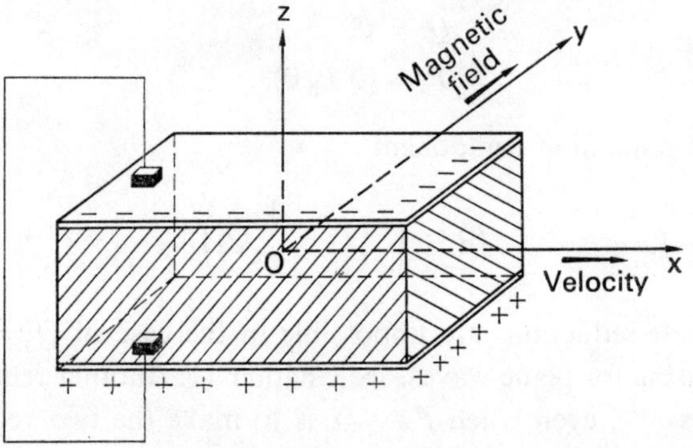

FIG. 12.1

that $\boldsymbol{B} = \{0\ B\ 0\}$. We have to apply the constitutive relations in the dielectric, and accordingly we must transform to a frame of reference in which the dielectric is at rest. In this frame of reference, because the plates are short-circuited, it follows that $E_3 = 0$ in the dielectric and therefore also, in particular,

$$H_2' = \beta(H_2+VD_3),$$
$$B_2' = \beta B_2,$$
$$E_3' = \beta V B_2,$$
$$D_3' = \beta\left(D_3+\frac{V}{c^2}H_2\right).$$

Applying the constitutive relations gives

$$B_2 = \mu(H_2+VD_3), \quad D_3+\frac{V}{c^2}H_2 = \varepsilon V B_2,$$

or, as an expression for $\boldsymbol{D}$ in terms of the magnetic field,

$$D_3 = \frac{VH_2(\mu\varepsilon-1/c^2)}{1-\mu\varepsilon V^2}.$$

The way in which the experiment is carried out is for the direction of the magnetic field to be suddenly reversed, in which case a charge is observed to flow by means of the ballistic galvanometer. This charge moves because the value of D_3 alters by an amount

$$\frac{2VH_2(\mu\varepsilon - 1/c^2)}{1-\mu\varepsilon V^2},$$

which is equal to the change in σ, the charge density on the plates. The magnitude of this charge is in good agreement with the results predicted by the theory. The experiment is difficult to discuss at all without special relativity, but the most plausible classical argument gives a corresponding result with μ replaced by μ_0 (so that the magnetic properties of the dielectric should not affect the result according to classical theory).

12.5 The electromagnetic potentials

It remains to say a few words about the representation of the field vectors by means of scalar and vector potentials. We saw above that, since $\left(\frac{1}{c^2}\frac{\partial}{\partial t} - \nabla\right)$, is a four-vector, any four-vector $(\phi, \boldsymbol{A})$ gives rise to

(i) an invariant $\quad \frac{\partial \phi}{\partial t} + \text{div}\, \boldsymbol{A},$

(ii) a six-vector $\quad \left\{-\frac{1}{c^2}\left(\frac{\partial \boldsymbol{A}}{\partial t} + \nabla(\phi c^2)\right), \ \textbf{curl}\, \boldsymbol{A}\right\}.$

Let us now compare these expressions with those derived in Volume 2 for Maxwell's equations

$$\textbf{curl}\, \boldsymbol{E} = -\frac{\partial \boldsymbol{B}}{\partial t}, \qquad \text{div}\, \boldsymbol{B} = 0.$$

From div $\boldsymbol{B} = 0$ it follows that $\boldsymbol{B} = \textbf{curl}\, \boldsymbol{A}$ and so the remaining equation becomes

$$\textbf{curl}\left(\boldsymbol{E} + \frac{\partial \boldsymbol{A}}{\partial t}\right) = 0,$$

showing that $\boldsymbol{E} + \partial \boldsymbol{A}/\partial t$ is a gradient. This suggests that the six-vector (ii) is $(\boldsymbol{E}/c^2, \boldsymbol{B})$, and that

$$\boldsymbol{E} = -\frac{\partial \boldsymbol{A}}{\partial t} - c^2\, \nabla \phi.$$

In other words, the time-component of the four-vector $(\phi, \boldsymbol{A})$ is not the usual electrostatic potential but differs from it by a factor c^2. If instead V has its usual meaning in electrostatics, then $(V/c^2, \boldsymbol{A})$ is a four-vector. This identification is confirmed by observing that the invariant expression (i) is the one expected from the usual Lorentz condition [eqn. 13.5].

Example 1. If $(\boldsymbol{P}, \boldsymbol{Q})$ is a six-vector, show that $(\boldsymbol{Q}, -c^2\boldsymbol{P})$ is a six-vector also, and therefore show that the fact that $(\boldsymbol{E}/c^2, \boldsymbol{B})$ is a six-vector is consistent with the fact (found earlier) that $(\boldsymbol{B}, -\boldsymbol{E})$ is a six-vector.

We can rewrite the transformation equations for $\boldsymbol{P}$ in the form

$$-c^2P_1' = -c^2P_1,$$
$$-c^2P_2' = \beta(-c^2P_2 + VQ_3),$$
$$-c^2P_3' = \beta(-c^2P_3 - VQ_2),$$

and similarly for those of Q.

Example 2. Show that Maxwell's equations for a field in empty space are covariant under Lorentz transformations.

The equations $\boldsymbol{D} = \varepsilon\boldsymbol{E}$ and $\boldsymbol{B} = \mu\boldsymbol{H}$ are valid for a medium at rest. Obtain expressions for $\boldsymbol{D}$ and $\boldsymbol{B}$ in terms of $\boldsymbol{E}$, $\boldsymbol{H}$, ε, μ when the medium is moving with uniform velocity $\boldsymbol{v}$.

Choose $\boldsymbol{v}$ in the direction of the x-axis. We then have

$$D_1' = D_1 = \varepsilon E_1 = \varepsilon E_1',$$
$$D_2' = \beta\left(D_2 - \frac{v}{c^2}H_3\right)$$
$$= \beta\left(\varepsilon E_2 - \frac{v}{c^2}H_3\right)$$
$$= \beta^2\varepsilon(E_2' + vB_3') - \frac{\beta^2 v}{c^2}(H_3' + vD_2'),$$

so that

$$D_2' = \varepsilon E_2'\varepsilon vB_3' - \frac{v}{c^2}H_3'.$$

Similarly

$$B_3' = \beta\left(B_3 - \frac{v}{c^2}E_2\right)$$
$$= \beta\left(\mu H_3 - \frac{v}{c^2}E_2\right)$$
$$= \beta^2\left\{\mu H_3' + \mu vD_2' - \frac{v}{c^2}(E_2' + vB_3')\right\},$$

which gives

$$B_3' = \mu H_3' + \mu vD_2' - \frac{v}{c^2}E_2'.$$

Eliminating D_2' gives

$$(1-\mu\varepsilon v^2)B_3' = \left(\mu\varepsilon-\frac{1}{c^2}\right)vE_2'+\mu\left(1-\frac{v^2}{c^2}\right)H_3'.$$

Similarly for the other components

$$D_3' = \varepsilon E_3'-\varepsilon vB_3'+(v/c^2)H_2',$$
$$B_2' = \mu H_3'-\mu vD_3'+(v/c^2)E_3'.$$

Example 3. A linearly polarized plane electromagnetic wave is propagated in free space from a transmitter fixed in an inertial frame S. The fields of the transmitted wave observed in S are

$$\boldsymbol{E} = \left\{0\ A_i \exp \mathrm{i}\omega_i\left(t-\frac{x}{c}\right) 0\right\}$$

$$\boldsymbol{B} = \left\{0\ 0\ c^{-1}A_i \exp \mathrm{i}\omega_i\left(t-\frac{x}{c}\right)\right\}.$$

The wave is normally incident on the plane face $x' = 0$ of a dielectric medium of refractive index n carried by a frame S' moving with uniform velocity $\{V\ 0\ 0\}$ relative to S. Prove that

(i) in the frame S', the observed reflection coefficient is $(1-n)/(1+n)$;
(ii) in the frame S, the observed frequency ω_r of the reflected wave is given by

$$\omega_r = \omega_i\left(1-\frac{V}{c}\right)\Big/\left(1+\frac{V}{c}\right);$$

(iii) in the frame S the observed reflection coefficient is

$$\left(1-\frac{V}{c}\right)(1-n)\Big/\left(1+\frac{V}{c}\right)(1+n).$$

In the dielectric the speed of propagation of light is c/n (see Fig. 12.2).

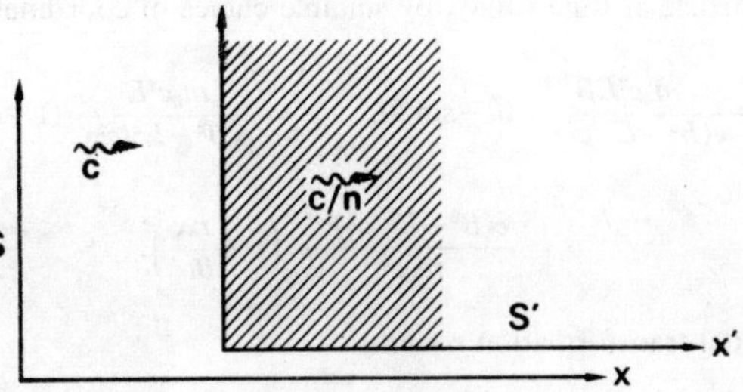

FIG. 12.2

(i) The observed reflection coefficient in S' is that for the wave normally incident on a stationary plane face; this is the type of problem on plane waves considered in Chapter 11, where the reflection coefficient was shown to be $(1-n)/(1+n)$, see p. 447.

(ii) In S' the fields are

$$\boldsymbol{E}' = \{0\ \beta(1-v/c)\,A_i\,\mathrm{e}^{\mathrm{i}\omega_i(t-x/c)}\ 0\},$$

$$\boldsymbol{B}' = \left\{0\ 0\ \frac{1}{c}\,\beta(1-v/c)\,A_i\,\mathrm{e}^{\mathrm{i}\omega_i(t-x/c)}\right\}.$$

Moreover, the indices of the exponential are such that

$$\omega_i(t-x/c) = \omega_i'(t'-x'/c)$$
$$= \omega_i'\beta(t-x/c)(1+v/c),$$
$$\omega_i = \omega_i'\sqrt{\left(\frac{1+v/c}{1-v/c}\right)}.$$

But in S' the face is stationary, so the reflected frequency is

$$\omega_r' = \omega_i' = \omega_i\sqrt{\left(\frac{1-v/c}{1+v/c}\right)}.$$

The exponent of the reflected wave is therefore

$$\mathrm{i}\omega_i\sqrt{\left(\frac{1-v/c}{1+v/c}\right)}(t'+x'/c)$$
$$= \mathrm{i}\omega_i\sqrt{\left(\frac{1-v/c}{1+v/c}\right)}(t+x/c)(1-v/c)$$
$$= \mathrm{i}\omega_i\sqrt{\left(\frac{1-v/c}{1+v/c}\right)}(t+x/c)$$
$$= \mathrm{i}\omega_i'(t+x/c).$$

Hence the reflected frequency is as stated.

(iii) The reflection coefficient is found in a corresponding way.

Example 4. A particle of rest mass m_0 and charge e moves from rest in a uniform electric field $E\boldsymbol{j}$ and a uniform magnetic field $B\boldsymbol{k}$, where $E < cB$. By means of a moving frame of reference such as will eliminate the electric field, or otherwise, show that the position of the particle at time t may, by suitable choice of coordinates, be expressed in the form

$$x = \frac{m_0c^2EB}{e(B^2-E^2/c^2)^{3/2}}(\theta-\sin\theta), \quad y = \frac{m_0c^2E}{e(B^2-E^2/c^2)}(1-\cos\theta)$$

where

$$\theta = \frac{e(B^2-E^2/c^2)^{1/2}}{m_0c^2}\left(t-\frac{Ex}{Bc^2}\right).$$

Making the usual transformation we find

$$E_2' = \beta(E-VB) = 0 \quad \text{if} \quad V = E/B < c.$$

Then also

$$B_3' = \beta\left(B-\frac{E^2}{Bc^2}\right) = \frac{\beta}{Bc^2}\left(B^2-\frac{E^2}{c^2}\right)$$
$$= \frac{B}{\sqrt{(B^2-E^2/c^2)}}\frac{1}{Bc^2}(B^2-E^2/c^2) = \frac{1}{c^2}\sqrt{(B^2-E^2/c^2)}.$$

Hence $$\frac{\mathrm{d}}{\mathrm{d}t'}(m\dot{\boldsymbol{r}}') = e(\boldsymbol{E}'+\boldsymbol{v}'\times\boldsymbol{B}'),$$

where $m = m_0/\sqrt{(1-v'^2/c^2)}$, so that

$$\frac{d}{dt'}(m\dot{z}') = 0, \quad \text{so that} \quad \dot{z}' = 0, \quad z' = 0 \text{ always},$$

$$\frac{d}{dt'}(m\dot{x}') = \frac{e}{c^2}\dot{y}'\sqrt{(B^2-E^2/c^2)},$$

$$\frac{d}{dt'}(m\dot{y}') = -\frac{e}{c^2}\dot{x}'\sqrt{(B^2-E^2/c^2)}.$$

From the last two equations

$$\dot{x}'\frac{d}{dt'}(m\dot{x}')+\dot{y}'\frac{d}{dt'}(m\dot{y}') = 0,$$

i.e.
$$v'^2\frac{dm}{dt'}+m\frac{d}{dt}\left(\frac{1}{2}v'^2\right) = 0.$$

Since this is easily seen to be inconsistent with the usual relation between m and v' it follows that $dm/dt' = 0$ and $v' =$ constant.

The starting velocity in this frame is $-V = -E/B$, so that this is the constant value of the velocity and

$$m = \frac{m_0}{\sqrt{\left(1-\frac{E^2}{c^2B^2}\right)}} = \frac{m_0B}{\sqrt{\left(B^2-\frac{E^2}{c^2}\right)}}.$$

Altogether, then,

$$m_0B\ddot{x}' = \frac{e}{c^2}\dot{y}'(B^2-E^2/c^2),$$

$$m_0B\ddot{y}' = -\frac{e}{c^2}\dot{x}'(B^2-E^2/c^2),$$

which integrate to give

$$\dot{x}'+i\dot{y}'+\frac{ie(B^2-E^2/c^2)}{m_0c^2B}(x'+iy') = -\frac{E}{B},$$

i.e.
$$\frac{d}{dt'}[(x'+iy')\exp\{ie(B^2-E^2/c^2)t'/(m_0c^2B)\}] = -\frac{E}{B}[\exp\{ie(B^2-E^2/c^2)t'/(m_0c^2B)\}]$$

so that

$$(x'+iy')\exp\{ie(B^2-E^2/c^2)t'/(m_0c^2B)\} = -\frac{Em_0c^2}{ie(B^2-E^2/c^2)}[\exp\{ie(B^2-E^2/c^2)t'/(m_0c^2B)\}-1].$$

But

$$t' = \beta\left(t-\frac{vx}{c^2}\right) = \frac{B}{\sqrt{(B^2-E^2/c^2)}}\left(t-\frac{Ex}{Bc^2}\right),$$

so the exponents are θ, and

$$x'+iy' = -\frac{Em_0c^2}{ie(B^2-E^2/c^2)}\{1-e^{-i\theta}\}.$$

Hence

$$x' = -\frac{Em_0c^2}{e(B^2-E^2/c^2)}\sin\theta,$$

$$y' = +\frac{Em_0c^2}{e(B^2-E^2/c^2)}(1-\cos\theta) = y.$$

The expression for x comes from

$$x = \beta(x'+vt')$$
$$= \frac{B}{\sqrt{(B^2-E^2/c^2)}}\left\{\frac{E}{B}\frac{B}{\sqrt{(B^2-E^2/c^2)}}\left(t-\frac{Ex}{Bc^2}\right)-\frac{Em_0c^2}{e(B^2-E^2/c^2)}\sin\theta\right\}$$
$$= \frac{m_0c^2EB}{e(B^2-E^2/c^2)^{3/2}}\{\theta-\sin\theta\}.$$

Miscellaneous Exercises XII.

1. (*a*) A particle of charge e and mass m is in a region of space in which there are uniform electric and magnetic fields perpendicular to each other. Find the possible states of motion of the particle for which the acceleration will be zero.

(*b*) Prove that if $\boldsymbol{E}$ and $\boldsymbol{B}$ are perpendicular in one Lorentz frame, they are perpendicular in all Lorentz frames, and that if $|\boldsymbol{E}|<|c\boldsymbol{B}|$ in one Lorentz frame, then $|\boldsymbol{E}|<|c\boldsymbol{B}|$ in all Lorentz frames.

2. Prove by direct use of the Lorentz transformation that the operator $\nabla^2-(1/c^2)(\partial^2/\partial t^2)$ is invariant under the transformation.

If $\boldsymbol{j}$, ϱ, $\boldsymbol{A}$, ϕ are the usual current-density vector, charge density, vector potential and scalar potential, so that

$$\boldsymbol{E} = -\nabla\phi-\frac{\partial \boldsymbol{A}}{\partial t},$$
$$\boldsymbol{B} = \text{curl}\,\boldsymbol{A},$$
$$\text{curl}\,\boldsymbol{H} = \frac{\partial \boldsymbol{D}}{\partial t}+\boldsymbol{j},$$
$$\text{div}\,\boldsymbol{D} = \varrho,$$

show that (ϱ, j_1, j_2, j_3) is a four-vector, and so is $(\phi c^2, A_1, A_2, A_3)$.

3. A classical point magnetic dipole $\boldsymbol{\mu}$ at rest has a vector potential $\boldsymbol{A} = \boldsymbol{\mu}\times\boldsymbol{r}/r^3$. Show that, if the magnetic dipole moves with a velocity $\boldsymbol{v}$ such that $v \ll c$, there is an electric dipole of moment $\boldsymbol{p}$ associated with the magnetic dipole, where $\boldsymbol{p} = \boldsymbol{v}\times\boldsymbol{\mu}$.

4. The equations for the electromagnetic four vector *in vacuo* are

$$\frac{\partial A_p}{\partial x_p} = 0, \qquad \frac{\partial^2 A_q}{\partial x_p^2} = 0 \qquad (p, q = 1, 2, 3, 4),$$

$$x_1 = x,\ x_2 = y,\ x_3 = z,\ x_4 = ict.$$

Verify that $A_p = a_p \exp(ik_qx_q)$ satisfies these equations, provided a_p, $k_q(p, q = 1, 2, 3, 4)$ are constants such that $a_pk_p = 0$, $k_p^2 = 0$. By considering the four-vector property of A_p, deduce that a_p must transform as a four-vector under the Lorentz transformation and that k_px_p must be a scalar, so that k_p also transforms as a four-vector.

An observer, moving with uniform velocity v in the negative x_1-direction, uses coordinates $\bar{x}_p$ related to x_p by

$$\bar{x}_1 = \frac{x_1-i(v/c)\,x_4}{\{1-(v^2/c^2)\}^{1/2}}, \qquad \bar{x}_2 = x_2, \qquad \bar{x}_3 = x_3, \qquad x_4 = \frac{x_4+i(v/c)\,x_1}{\{1-(v^2/c^2)\}^{1/2}}.$$

By considering a similar transformation for k_p show that the new observer regards the frequency of a plane monochromatic electromagnetic wave as being increased by the Doppler factor

$$\left(1+\frac{v\cos\alpha}{c}\right)\left(1-\frac{v^2}{c^2}\right)^{1/2},$$

where α is the angle between the direction of propagation and the positive x_1-direction as determined by the first observer. Show also that the corresponding angle $\bar{\alpha}$ determined by the new observer is given by

$$\cos\bar{\alpha} = \left(\cos\alpha+\frac{v}{c}\right)\left(1+\frac{v\cos\alpha}{c}\right)^{-1}.$$

5. Describe briefly a method for transforming the components of the electromagnetic field in free space from one Lorentz frame to another.

By means of this transformation, show that the field at a point P due to a charge e moving with uniform velocity $\boldsymbol{V}$ is, at time t,

$$\boldsymbol{E} = -\frac{e}{\beta^2}\left[\left(\boldsymbol{r}+\frac{r\boldsymbol{V}}{c}\right)\Big/ s^3\right], \quad \boldsymbol{H} = \boldsymbol{E}\times[\hat{\boldsymbol{r}}],$$

where $\beta = \beta(V)$, $\boldsymbol{r}$ is the position vector of the charge relative to P, $s = r+\boldsymbol{V}\cdot\boldsymbol{r}/c$, and square brackets indicate that the quantities enclosed are evaluated at time $t-[r]/c$. Show that the fields are as expected for $V \ll c$.

CHAPTER 13

RADIATION

13.1 General properties of radiation

We shall be interested in this chapter in electromagnetic fields which change at high frequency. In this case the inequality

$$\frac{\partial \boldsymbol{D}}{\partial t} \gg \boldsymbol{j}$$

holds. Accordingly the ordinary current in the Maxwell equations is negligible compared with the displacement current and we may take the equations as

$$\mathbf{curl}\, \boldsymbol{H} = \frac{\partial \boldsymbol{D}}{\partial t}, \qquad \mathbf{curl}\, \boldsymbol{E} = -\frac{\partial \boldsymbol{B}}{\partial t}$$

so that, since

$$\boldsymbol{B} = \mu_0 \boldsymbol{H}, \qquad \boldsymbol{D} = \varepsilon_0 \boldsymbol{E},$$

$$\mathbf{curl}\, \boldsymbol{H} = \varepsilon_0 \frac{\partial \boldsymbol{E}}{\partial t}, \qquad \mathbf{curl}\, \boldsymbol{E} = -\mu_0 \frac{\partial \boldsymbol{H}}{\partial t}.$$

By taking **curl** of each equation and using the identity

$$\mathbf{curl}\, \mathbf{curl}\, \boldsymbol{A} = \mathbf{grad}\, \mathrm{div}\, \boldsymbol{A} - \nabla^2 \boldsymbol{A},$$

one can at once derive

$$\frac{1}{c^2}\frac{\partial^2 \boldsymbol{E}}{\partial t^2} = \nabla^2 \boldsymbol{E}, \qquad \frac{1}{c^2}\frac{\partial^2 \boldsymbol{H}}{\partial t^2} = \nabla^2 \boldsymbol{H}, \tag{13.1}$$

where c is the velocity of propagation of the disturbance described by the equations and has the value $c = (\varepsilon_0 \mu_0)^{-1/2}$.

These equations contain the essential features of the propagation properties of the field and were investigated in detail by Hertz. His investigation will be given in the next section but before this it is a good idea to try and obtain some intuitive idea of how outgoing radiation behaves. It is clear from the equations, since they are wave equations, that the field will be transmitted with a finite velocity c. If one considers some particular changing configuration of charges, for example, the field of two equal and opposite charges, which give the field of a dipole at large distances, one can think of their electric and magnetic lines of force related to each other in a spatial pattern which moves with this velocity (of value about 3×10^8 m s^{-1}). In a complete oscillation of the charges the field lines will alter their direction twice, once in each half-period of oscillation, and so the change of direction becomes more rapid the higher the frequency of the oscillation. One can see in a general way the effect of this. Since the field propagates with a finite speed, the field lines which are remote from the source will have no time to return to the directions corresponding to the ones nearer to the charges. In other words, as frequency increases, the field lines tend to separate. Those which are near enough to the dipole will move away from it and back towards it, but there will be some critical surface which separates these from the more distant ones which cannot get back. These more distant ones correspond to the radiation field in which we are principally interested here.

13.2 The Hertz vector

In considering the fourth Maxwell equation

$$\operatorname{div} \boldsymbol{B} = 0$$

we have in earlier chapters made the substitution

$$\boldsymbol{B} = \textbf{curl}\, \boldsymbol{A}.$$

However, in this substitution it is clear that the vector $\boldsymbol{A}$ is not uniquely determined and could be altered to the form

$$\boldsymbol{A}' = \boldsymbol{A} + \nabla\psi,$$

and accordingly further restrictions are usually placed upon the vector potential $\boldsymbol{A}$. In the case of the static or slowly changing field the most convenient further restriction is

$$\operatorname{div} \boldsymbol{A}' = 0. \tag{13.2}$$

This involves solving the equation

$$\nabla^2\psi = -\operatorname{div} \boldsymbol{A},$$

Poisson's equation, and this is always possible in general. However, it will prove to be more satisfactory to use a different restriction here. Substituting in the equation

$$\boldsymbol{H} = \frac{1}{\mu_0}\,\mathbf{curl}\,\boldsymbol{A} \tag{13.3}$$

we find

$$\mathbf{curl}\left(\boldsymbol{E}+\frac{\partial \boldsymbol{A}}{\partial t}\right) = \boldsymbol{0}.$$

From this it follows in the usual way that

$$\boldsymbol{E} = -\nabla\phi - \frac{\partial \boldsymbol{A}}{\partial t}, \tag{13.4}$$

and by putting this back into the original equations and using the identity for the repeated curl again, the equation

$$\mathbf{grad}\,\mathrm{div}\,\boldsymbol{A} - \nabla^2\boldsymbol{A} = -\mu_0\varepsilon_0\left(\mathbf{grad}\,\frac{\partial \phi}{\partial t} + \frac{\partial^2 \boldsymbol{A}}{\partial t^2}\right)$$

results. It would be very convenient at this point if the vector $\boldsymbol{A}$ also satisfied the wave equation. We see that this is indeed the case provided that

$$\nabla\left(\mathrm{div}\,\boldsymbol{A} + \varepsilon_0\mu_0\frac{\partial \phi}{\partial t}\right) = \boldsymbol{0}.$$

Since we have the possibility of imposing some additional restriction, we can choose the restriction

$$\mathrm{div}\,\boldsymbol{A} + \frac{1}{c^2}\frac{\partial \phi}{\partial t} = 0.$$

If this condition is not satisfied already we can transform to a new vector potential $\boldsymbol{A}' = \boldsymbol{A} + \nabla\psi$. In order that $\boldsymbol{E}$ should be unchanged we must then transform to a new ϕ', so that

$$\nabla\phi' + \frac{\partial \boldsymbol{A}'}{\partial t} = \nabla\phi + \frac{\partial \boldsymbol{A}}{\partial t},$$

i.e.

$$\nabla(\phi' + \dot{\psi} - \phi) = \boldsymbol{0}.$$

Let us choose, then,

$$\phi' = \phi - \frac{\partial \psi}{\partial t},$$

$$\boldsymbol{A}' = \boldsymbol{A} + \nabla\psi,$$

and then

$$\operatorname{div} \boldsymbol{A}'+\frac{1}{c^2}\frac{\partial\phi'}{\partial t} = \left(\operatorname{div} \boldsymbol{A}+\frac{1}{c^2}\frac{\partial\phi}{\partial t}\right)-\left(\frac{1}{c^2}\frac{\partial^2\psi}{\partial t^2}-\nabla^2\psi\right),$$

so that we have only to solve the inhomogeneous wave equation

$$\left(\frac{1}{c^2}\frac{\partial^2}{\partial t^2}-\nabla^2\right)\psi = \operatorname{div} \boldsymbol{A}+\frac{1}{c^2}\frac{\partial\phi}{\partial t}.$$

The condition

$$\operatorname{div} \boldsymbol{A}+\frac{1}{c^2}\frac{\partial\phi}{\partial t} = 0 \tag{13.5}$$

is often known as the *Lorentz condition.* It has the effect of making both $\boldsymbol{A}$ and ϕ satisfy the same wave equations as the field vectors (which the reader may verify in the case of ϕ). Now the starting-point for Hertz's analysis is to notice that, if the potentials are to satisfy the Lorentz condition, then they can always be written in the form

$$\phi = -\operatorname{div} \boldsymbol{\Pi}, \qquad \boldsymbol{A} = \frac{1}{c^2}\frac{\partial\boldsymbol{\Pi}}{\partial t}, \tag{13.6}$$

where $\boldsymbol{\Pi}$ is some new vector, and it is *again* the case that the new vector which has been introduced here, known as the *Hertz potential*, satisfies the two conditions

$$\frac{\partial}{\partial t}\left(\frac{1}{c^2}\frac{\partial^2\boldsymbol{\Pi}}{\partial t^2}-\nabla^2\boldsymbol{\Pi}\right) = \boldsymbol{0},$$

$$\operatorname{div}\left(\frac{1}{c^2}\frac{\partial^2\boldsymbol{\Pi}}{\partial t^2}-\nabla^2\boldsymbol{\Pi}\right) = 0.$$

These conditions are very reminiscent of the wave equation and accordingly we propose to consider the special case in which the field satisfies

$$\frac{1}{c^2}\frac{\partial^2\boldsymbol{\Pi}}{\partial t^2}-\nabla^2\boldsymbol{\Pi} = \boldsymbol{0}. \tag{13.7}$$

(This was the special case considered by Hertz.) Substituting for the potentials the field strengths take the form

$$\boldsymbol{H} = \varepsilon_0\frac{\partial}{\partial t}\,\mathbf{curl}\,\boldsymbol{\Pi}, \tag{13.8}$$

$$\boldsymbol{E} = \mathbf{grad}\operatorname{div}\boldsymbol{\Pi}-\frac{1}{c^2}\frac{\partial^2\boldsymbol{\Pi}}{\partial t^2} = \mathbf{curl}\,\mathbf{curl}\,\boldsymbol{\Pi}. \tag{13.9}$$

Exercises 13.2

1. A region contains no charge or current densities but contains a distribution of electric polarization $\boldsymbol{P}$, so that $\boldsymbol{D} = \varepsilon\boldsymbol{E}+\boldsymbol{P}$, $\boldsymbol{B} = \mu\boldsymbol{H}$. Show that the Hertz potential satisfies the equations

$$\operatorname{div}\left(\nabla^2\boldsymbol{\Pi} - \mu\varepsilon\frac{\partial^2\boldsymbol{\Pi}}{\partial t^2} + \frac{\boldsymbol{P}}{\varepsilon}\right) = 0, \quad \frac{\partial}{\partial t}\left(\nabla^2\boldsymbol{\Pi} - \mu\varepsilon\frac{\partial^2\boldsymbol{\Pi}}{\partial t^2} + \frac{\boldsymbol{P}}{\varepsilon}\right) = \boldsymbol{0},$$

in the region concerned.
(This relates the Hertz potential to the distribution of electric dipoles.)

2. Show that Maxwell's equations, in the absence of charge and current densities, are satisfied by the following substitutions

$$\boldsymbol{D} = \textbf{curl}\,\boldsymbol{A}^*, \quad \boldsymbol{H} = \textbf{grad}\,V^* + \frac{\partial \boldsymbol{A}^*}{\partial t},$$

where

$$\nabla^2 V^* - \mu\varepsilon\frac{\partial^2 V^*}{\partial t} = 0, \quad \nabla^2\boldsymbol{A}^* - \mu\varepsilon\frac{\partial^2\boldsymbol{A}^*}{\partial t^2} = \boldsymbol{0},$$

if

$$\operatorname{div}\boldsymbol{A}^* + \mu\varepsilon\frac{\partial V^*}{\partial t} = 0.$$

3. A region contains no charge or current densities but contains a distribution of magnetic polarization so that $\boldsymbol{D} = \varepsilon\boldsymbol{E}$, $\boldsymbol{B} = \mu(\boldsymbol{H}+\boldsymbol{M})$. Show that Maxwell's equations are satisfied by (see eqn. 2)

$$\boldsymbol{D} = \textbf{curl}\,\boldsymbol{A}^*, \quad \boldsymbol{H} = \textbf{grad}\,V^* + \frac{\partial \boldsymbol{A}^*}{\partial t},$$

with

$$\boldsymbol{A}^* = -\mu\varepsilon\frac{\partial\boldsymbol{\Pi}^*}{\partial t}, \quad V^* = \operatorname{div}\boldsymbol{\Pi}^*$$

where

$$\operatorname{div}\left(\nabla^2\boldsymbol{\Pi}^* - \mu\varepsilon\frac{\partial^2\boldsymbol{\Pi}^*}{\partial t^2} + \boldsymbol{M}\right) = 0, \quad \frac{\partial}{\partial t}\left(\nabla^2\boldsymbol{\Pi}^* - \mu\varepsilon\frac{\partial^2\boldsymbol{\Pi}^*}{\partial t^2} + \boldsymbol{M}\right) = \boldsymbol{0}.$$

13.3 Solutions with axial symmetry

The use of Hertz's vector is particularly appropriate in conditions of symmetry. The first kind of symmetry which springs to mind is that of spherical symmetry. Now the Hertz vector would have the form

$$\boldsymbol{\Pi} = X\boldsymbol{r}$$

where $X = X(r, t)$ is a function of r and t only. However, when one prepares to substitute this value into the expression for the field strength, it at once follows that

$$\textbf{curl}\,\boldsymbol{\Pi} = (\nabla X)\times\boldsymbol{r} = \boldsymbol{0},$$

corresponding to zero field. This suggests that spherically symmetric waves are impossible, a conclusion which is immediately verified by making the substitution

$$\boldsymbol{E} = f(r)\boldsymbol{r}, \quad \boldsymbol{H} = g(r)\boldsymbol{r},$$

in the original equations. It is therefore necessary to go up to the next stage of complication, that is, to the situation where there is symmetry about a line, which we may take as the z-axis.

We shall suppose that the charges present, which produce the field, are located very near to the origin and are fluctuating in some way so as to produce the radiation conditions necessary. In fact the solution which we shall find is mainly applicable to a dipole whose strength varies with the time, but we shall also consider the general case. It is natural to use polar coordinates for the more detailed calculations; because the axis of symmetry is the z-axis, the angle between this and the radius vector r may be taken as θ. The Hertz vector, $\boldsymbol{\Pi}$, in order to preserve the axial symmetry can only have components radially outwards and parallel to the z-axis. Both of these components can only depend upon the distance from the origin. However, Maxwell's equations are all linear equations, so that the radial component which would correspond to a superimposed spherically symmetrical field will in fact contribute nothing and we may choose for the Hertz vector some vector in the direction of the z-axis.

This Hertz vector now satisfies the wave equation which may be expressed in polar coordinates as

$$\frac{\partial^2 \boldsymbol{\Pi}}{\partial r^2} + \frac{2}{r}\frac{\partial \boldsymbol{\Pi}}{\partial r} = \frac{1}{c^2}\frac{\partial^2 \boldsymbol{\Pi}}{\partial t^2}. \tag{13.10}$$

In this equation the Hertz vector is to be assumed in the form

$$\boldsymbol{\Pi} = f(r, t)\boldsymbol{k}, \tag{13.11}$$

so that we have, in effect, a scalar wave equation for the function f. We shall be particularly interested in the special case in which the wave field has a periodic character, in which case we would write the Hertz vector in the simpler form

$$\boldsymbol{\Pi} = \Pi(r)e^{i\omega t}\boldsymbol{k}. \tag{13.12}$$

(Here, as usual, the real part of the exponential function is intended.) By making this substitution in the wave equation we obtain the ordinary differential equation for the dependence of Π on the distance, viz.

$$\frac{\partial^2 \Pi}{\partial r^2} + \frac{2}{r}\frac{\partial \Pi}{\partial r} + \frac{\omega^2}{c^2}\Pi = 0.$$

This equation may be integrated at once by the substitution $r\Pi = u$ and leads to the solution

$$\Pi(r) = \frac{\Pi_0}{r} e^{-i\omega r/c}.$$

The Hertz vector therefore takes the form, in this particular case,

$$\boldsymbol{\Pi} = \frac{\Pi_0}{r} \exp\{i\omega(t-r/c)\}\boldsymbol{k} = f(r, t)\boldsymbol{k}, \tag{13.13}$$

in which the outgoing wave nature of the solution is clearly exhibited. In the general case the function f must simply be some solution of the wave equation and therefore has the form

$$\boldsymbol{\Pi} = \frac{1}{r} g(t-r/c)\boldsymbol{k}, \tag{13.14}$$

by means of a similar argument. However, it will be more convenient to leave it in its original form $\boldsymbol{\Pi} = f(r, t)\boldsymbol{k}$ for subsequent calculations.

13.4 Discussion of the field strength

Taking then the Hertz vector in the form

$$\boldsymbol{\Pi} = f(r, t)\boldsymbol{k},$$

it follows that

$$\textbf{curl}\, \boldsymbol{\Pi} = \frac{\partial f}{\partial r} \hat{\boldsymbol{r}} \times \boldsymbol{k}.$$

Accordingly the magnetic field is given by

$$\boldsymbol{H} = \varepsilon_0 \frac{\partial^2 f}{\partial r\, \partial t} \hat{\boldsymbol{r}} \times \boldsymbol{k}, \tag{13.15}$$

and so the magnetic field lines are circles whose centres lie on the z-axis. In polar coordinates the only non-zero component of the magnetic field is therefore

$$H_\phi = \varepsilon_0 \frac{\partial^2 f}{\partial r\, \partial t} \sin\theta.$$

Taking the particular case

$$f(r, t) = \frac{M}{4\pi\varepsilon_0 r} \exp\{i\omega(t-r/c)\},$$

where $M/(4\pi\varepsilon_0)$ is now written for the constant Π_0, corresponding to periodic solutions, an immediate calculation gives

$$H_\phi = \frac{M}{4\pi}\sin\theta\left\{-\frac{i\omega}{r^2}+\frac{\omega^2}{cr}\right\}\exp\{i\omega(t-r/c)\}.$$

Since the real part has to be taken, this means that the component of the magnetic field has the form

$$H_\phi = \frac{M}{4\pi}\sin\theta\left\{\frac{\omega}{r^2}\sin\omega\left(t-\frac{r}{c}\right)+\frac{\omega^2}{cr}\cos\omega\left(t-\frac{r}{c}\right)\right\}. \tag{13.16}$$

We notice that one of these parts falls off inversely as the square of the distance while the other falls off inversely as the distance. The distinction between these two parts corresponds to that between the near zone and the distant zone mentioned in our earlier intuitive argument.

Next, for the calculation of the electric field we have

$$\boldsymbol{E} = \textbf{curl}\left\{\frac{1}{r}\frac{\partial f}{\partial r}\hat{\boldsymbol{r}}\times\boldsymbol{k}\right\},$$

and this immediately reduces to the form

$$\boldsymbol{E} = -\frac{2}{r}\frac{\partial f}{\partial r}\boldsymbol{k}+r\frac{\partial}{\partial r}\left(\frac{1}{r}\frac{\partial f}{\partial r}\right)(\hat{\boldsymbol{r}}\cdot\boldsymbol{k}\hat{\boldsymbol{r}}-\boldsymbol{k}).$$

The electric field now has components in the two directions at right angles to the magnetic field, radially and transversely. The value of these components can at once be calculated as

$$E_r = \hat{\boldsymbol{r}}\cdot\boldsymbol{E} = -\frac{2}{r}\frac{\partial f}{\partial r}\cos\theta,$$

$$E_\theta = \frac{2}{r}\frac{\partial f}{\partial r}\sin\theta+r\frac{\partial}{\partial r}\left(\frac{1}{r}\frac{\partial f}{\partial r}\right)\sin\theta.$$

Taking again the particular case of the periodic solution, these components have the values

$$E_r = \frac{M\cos\theta}{2\pi\varepsilon_0}\left\{\frac{1}{r^3}\cos\omega\left(t-\frac{r}{c}\right)-\frac{\omega}{cr^2}\sin\omega\left(t-\frac{r}{c}\right)\right\}, \tag{13.17}$$

$$E_\theta = \frac{M\sin\theta}{4\pi\varepsilon_0}\left\{\frac{1}{r^3}\cos\omega\left(t-\frac{r}{c}\right)-\frac{\omega}{cr^2}\sin\omega\left(t-\frac{r}{c}\right)-\frac{\omega^2}{c^2r}\cos\omega\left(t-\frac{r}{c}\right)\right\}. \tag{13.18}$$

Here again the distinction between the near and distant zones is very clear. The part of the electric field which is important nearest to the origin falls off as the cube of the distance and the part which is important a great way off again falls off inversely as the distance. Approximately, one can say that the values of the fields near to the oscillating system are

$$E_r = \frac{M \cos \theta}{2\pi\varepsilon_0 r^3} \cos \omega\left(t-\frac{r}{c}\right), \tag{13.19}$$

$$E_\theta = \frac{M \sin \theta}{4\pi\varepsilon_0 r^3} \cos \omega\left(t-\frac{r}{c}\right), \tag{13.20}$$

$$H_\phi = \frac{\omega M \sin \theta}{4\pi r^2} \sin \omega\left(t-\frac{r}{c}\right), \tag{13.21}$$

whereas those a great way away have the form

$$E_r = 0, \tag{13.22}$$

$$E_\theta = -\frac{M\omega^2 \sin \theta}{4\pi\varepsilon_0 c^2 r} \cos \omega\left(t-\frac{r}{c}\right), \tag{13.23}$$

$$H_\phi = \frac{M\omega^2 \sin \theta}{4\pi c r} \cos \omega\left(t-\frac{r}{c}\right), \tag{13.24}$$

approximately. It is important to notice that the field in the distant zone satisfies the two conditions

$$\varepsilon_0 \boldsymbol{E}^2 = \mu_0 \boldsymbol{H}^2, \quad \boldsymbol{E}\cdot\boldsymbol{H} = 0 \tag{13.25}$$

which we have already discussed in connection with plane radiation fields. These conditions do not depend upon the particular Lorentz frame of reference.

13.5 Interpretation of the results

We first consider the results corresponding to a periodic oscillation because the interpretation of this is most straightforward and this is in fact the most important form of solution. Very near to the oscillating charges the fields have the form

$$E_r = \frac{M \cos \theta}{2\pi\varepsilon_0 r^3} \cos \omega t,$$

$$E_\theta = \frac{M \sin \theta}{4\pi\varepsilon_0 r^3} \cos \omega t,$$

$$H_\phi = \frac{M\omega \sin \theta}{4\pi r^2} \sin \omega t$$

In order to interpret this one needs to compare the electric field with that of a dipole. The field of a dipole of moment m directed along the polar axis is

$$E_r = \frac{m \cos \theta}{2\pi\varepsilon_0 r^3}, \quad E_\theta = \frac{m \sin \theta}{4\pi\varepsilon_0 r^3},$$

so that the electric field corresponds to such a dipole with varying moment m. Now such a varying moment may be considered as due to a current element. One can think of a current element whose length is equal to l, that of the dipole (considered as a very short length). Accordingly the value of H for such an element will be given by

$$H_\phi = \frac{\sqrt{2}Il \sin \theta}{4\pi r^2} \sin \omega t$$

where I is the r.m.s. value of the current (so that $I\sqrt{2}$ is the peak value) and ω is its frequency. If now

$$Il\sqrt{2} = \omega M,$$

the magnetic fields agree. Moreover, this equation involves

$$M = \frac{Il\sqrt{2}}{\omega} = ql.$$

When the dipole is thought of as a charge oscillating up and down a short length l the charges at the ends of l are $\pm q \cos \omega t$ and $m = ql \cos \omega t$. The current I in the element arises from the variations given by $d/dt(q \cos \omega t) = -\omega q \sin \omega t$. Hence the peak value is $I\sqrt{2} = \omega q$. This is in complete agreement with the electrical picture. Thus both the fields imply that the solution we have found corresponds to an oscillating dipole at the origin directed up the z-axis.

It is to be noticed that from the equations for the field in the near zone the squares of the field strength (which enter into the expression for the energy) are inversely proportional to the sixth power of r. Moreover, the electric and magnetic field strength differ in phase by $\frac{1}{2}\pi$. The Poynting vector, which describes the passage of energy, is therefore proportional to (using the polar components):

$$\boldsymbol{E}\times\boldsymbol{H} = (E_\theta H_\phi, -E_r H_\phi, 0),$$

i.e. has both its components proportional to

$$\cos \omega\left(t-\frac{r}{c}\right) \sin \omega\left(t-\frac{r}{c}\right) = \frac{1}{2} \sin \left\{2\omega\left(t-\frac{r}{c}\right)\right\}.$$

As a result of this the energy from the dipole flows out during half the period of oscillation into the near zone, and then in the next half it flows back again. On the other hand, the equations for the distant zone show the electric and magnetic fields in phase. The Poynting vector now has components proportional to $\{E_\theta H_\phi \ 0 \ 0\}$, i.e. it is purely radial and proportional to $\cos^2\omega(t-r/c)$. There is therefore an energy-flow outwards. (Since the radial component of the electric field vanishes the shape of the distant field is generally a cylindrical one, which accounts for the fact that it falls off inversely as the distance.) As a result the energy decreases only as the inverse square of the distance.

Thus apart from the two zones differing by the size of the fields and the manner in which these fields fall off with distance there is also another important distinction in the behaviour of the energy in them. In the near zone the energy flows out and back, so that in so far as the transmission of radiation is concerned the whole behaviour is really a pretence. On the other hand, the energy in the distant zone flows continually outwards and so there is a genuine transmission of radiation. The region in which the pretence takes place will be larger when the near zone is larger, that is, the lower the frequency or the longer the wavelength. For a wavelength of about 60 cm the near zone has a radius of about 5 cm, whereas for a wavelength of 6 km the near zone has a radius of about 500 m. This is the basic reason why experimental demonstration of the waves which Maxwell saw must exist was delayed for about 10 years. It is also the reason why long-wave radio transmitters need a large amount of apparatus and a considerable source of energy whereas very short-wave transmitters may be small enough to be carried in the hand. The large amount of energy consumed by the long-wave transmitter is mostly used in setting up the pretence of transmission in the near zone which is in fact only a pumping in and out of energy.

13.6 Other kinds of radiative solutions

The solution of Hertz, which has been described above, is historically of the greatest importance, but the reader will have noticed that there is a certain element of luck in its derivation. Certain plausible assumptions about the scalar and vector potentials lead to a solution which can then, by looking at the near field, be seen to correspond to an oscillating dipole at the origin. Two different directions suggest themselves for proceeding from this point. Firstly we might preserve the general features of the derivation of the scalar and vector potentials from a Hertz vector which is again the product of a constant vector and a scalar field (but without making the same restrictions on the dependence of the scalar fields on time and space variables). In the present section we look at one example of a solution found in this way. Alternatively we might seek to find the radiation field from a system of moving charges, by building this field up from that of a single charge. In this way the physical significance of what is found is certain, but there is the disadvantage that the calculation of the field in general is not so easy.

Returning then to the first course of action, let us write

$$\boldsymbol{\Pi} = \boldsymbol{k}\, f(\boldsymbol{r}, t),$$

where $\boldsymbol{k}$ is a unit vector which may, without loss of generality, be chosen to lie along the z-axis, and f is now any function of position and time. Since

$$\phi = -\operatorname{div} \boldsymbol{\Pi}, \quad \boldsymbol{A} = \frac{1}{c^2}\frac{\partial \boldsymbol{\Pi}}{\partial t},$$

we have

$$\phi = -\boldsymbol{k}\cdot\nabla f, \quad \boldsymbol{A} = \frac{1}{c^2}\boldsymbol{k}\dot{f}. \tag{13.26}$$

From this it is easy to calculate the fields. Here f is a scalar function satisfying the wave equation

$$\frac{1}{c^2}\frac{\partial^2 f}{\partial t^2} - \nabla^2 f = 0.$$

From any solution of this equation we can derive corresponding radiation fields.

In illustration, instead of considering the type of solution involved in Hertz's calculation, let us try something of a more symmetrical kind. If f is independent of t, we know that there is a so-called elementary solution of the resultant equation (Laplace's equation) of the form $f = 1/r$. Now if, for the time being, we define $i\boldsymbol{r} = \boldsymbol{s}$ so that the wave equation becomes

$$\frac{\partial^2 f}{\partial (ct)^2} + \frac{\partial^2 f}{\partial s_1^2} + \frac{\partial^2 f}{\partial s_2^2} + \frac{\partial^2 f}{\partial s_3^2} = 0,$$

(i.e. Laplace's equation in *four* dimensions), we can expect a similar elementary solution except that the power of $R = \sqrt{(c^2t^2 + s_1^2 + s_2^2 + s_3^2)}$ may well be different. In fact, if

$$f = g(R),$$

then

$$\frac{\partial f}{\partial (ct)} = \frac{ct}{R} g',$$

so that

$$\frac{\partial^2 f}{\partial (ct)^2} = \frac{1}{R} g' + \frac{(ct)^2}{R^2} g'' - \frac{(ct)^2}{R^3} g'.$$

Hence $$\frac{\partial^2 f}{\partial (ct)^2}+\sum_{i=1}^{3}\frac{\partial^2 f}{\partial s_i^2}=\frac{3}{R}g'+g''=\frac{1}{R^3}(R^3g')'=0.$$

Therefore $$R^3g' = A \quad \text{(a constant)}$$

so that $$g=\frac{B}{R^2}+C, \tag{13.27}$$

where B, C are constants. Disregarding the constant C, since it results in no field, we can take the elementary solution as

$$g=\frac{1}{R^2}=\frac{1}{c^2t^2+s_1^2+s_2^2+s_3^2}=\frac{1}{c^2t^2-r^2}. \tag{13.28}$$

Strictly speaking, this is a solution of the wave equation at all points except those at which $c^2t^2 = r^2$, at which this solution has a singularity. If we confine ourselves to values of $t > 0$ we can think of this singularity as beginning at the origin at $t = 0$. Then, as t increases, the singular points all lie, at any instant, on a sphere of increasing radius (a sphere, in fact, whose velocity is given by $|\boldsymbol{r}|/t = c$). The solution therefore represents an extremely short pulse of light emitted at the origin at $t = 0$.

There is, however, a purely formal procedure that enables us to derive from such a solution, with a singular surface, another solution with no singularities. First suppose that the original singular event occurs, not at the origin, but at the time t_0 and the position $\boldsymbol{r}_0$. The solution then obviously becomes

$$g=\frac{1}{c^2(t-t_0)^2-(\boldsymbol{r}-\boldsymbol{r}_0)^2}. \tag{13.29}$$

In this solution one can replace t_0, $\boldsymbol{r}_0$ by *complex* numbers, say

$$t_0 = \alpha+i\beta,$$
$$\boldsymbol{r}_0 = \boldsymbol{a}+i\boldsymbol{b},$$

and a short calculation proves that here $[\beta \neq (1-v^2/c^2)^{-\frac{1}{2}}]$

$$g=\frac{1}{A-iB},$$

where $$A = c^2(t-\alpha)^2-(\boldsymbol{r}-\boldsymbol{a})^2-c^2\beta^2+\boldsymbol{b}^2,$$
$$B = 2[c^2\beta(t-\alpha)-\boldsymbol{b}\cdot(\boldsymbol{r}-\boldsymbol{a})].$$

Hence also

$$g=\frac{A+iB}{A^2+B^2}=U+iV$$

and, since g satisfies the wave equation, it follows that U, V satisfy it also. Now U, V can only be singular if $A^2+B^2 = 0$, and this involves $A = B = 0$. So long as $c^2\beta^2 > b^2$, this means that

$$c^2(t-\alpha)^2 > (\boldsymbol{r}-\boldsymbol{a})^2$$

(by using $A = 0$), and so

$$\begin{aligned}\beta^2c^2(t-\alpha)^2 &> \beta^2(\boldsymbol{r}-\boldsymbol{a})^2\\ &> \boldsymbol{b}^2(\boldsymbol{r}-\boldsymbol{a})^2/c^2\\ &\geqslant [\boldsymbol{b}\cdot(\boldsymbol{r}-\boldsymbol{a})]^2/c^2\end{aligned}$$

so that $B > 0$, and therefore U, V are non-singular.

Now that the complex numbers have done their work they can be disregarded; the linearity of the wave equation is what makes this method possible. Of course, by a (real) change of origin it is possible to make $\alpha = 0$, $\boldsymbol{a} = \boldsymbol{0}$. Since $c^2\beta^2 > \boldsymbol{b}^2$ it is also possible, by means of a real Lorentz transformation, to make the real four-vector $(\beta, \boldsymbol{b})$ correspond to the time-axis, i.e. to make $\boldsymbol{b} = \boldsymbol{0}$. Then

$$A = c^2t^2-\boldsymbol{r}^2-c^2\beta^2, \quad B = 2c^2\beta t,$$

and so the V solution (for example) has the form

$$V = \frac{2c^2\beta t}{(c^2t^2-\boldsymbol{r}^2-c^2\beta^2)^2+4c^4\beta^2t^2}. \tag{13.30}$$

How is one to interpret such a solution of Maxwell's equations? There are no singularities—that is, no sources of the field anywhere or at any time. But, none the less, there are field strengths that can be calculated with a little trouble from the potentials given by the formulae

$$\phi = -\boldsymbol{k}\cdot\nabla V, \quad \boldsymbol{A} = \frac{1}{c^2}\boldsymbol{k}\frac{\partial V}{\partial t}. \tag{13.31}$$

To get some idea of how the field varies, we can plot V as a function of time for given $\boldsymbol{r}$. The curve is obviously of the form shown in Fig. 13.1 since $V=0$

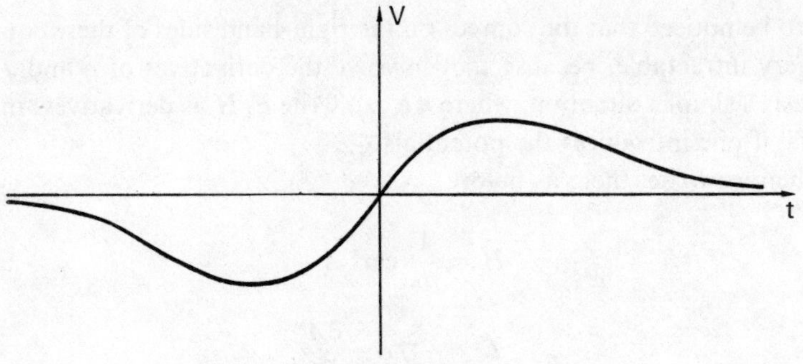

FIG. 13.1

when $t = 0$, $V \approx Kt$ when t is very small, and when t is large $V \approx K/t^3$. Thus a disturbance builds up, changes somewhat, and disappears again.

The existence of this kind of solution of Maxwell's equations suggests that Maxwell's theory may be incomplete. It seems to lack some additional restriction that will serve to ensure that fields originate only from sources like charges and magnets. But we do not know how to modify the theory so as to rectify this defect.

13.7 The fields of moving charges

We now seek to investigate the fields (in particular the radiation fields) of a number of charges in motion. Consider first the Maxwell equations in the absence of other matter than the charges:

$$\mathbf{curl}\,\boldsymbol{E} = -\mu_0 \frac{\partial \boldsymbol{H}}{\partial t}, \qquad \operatorname{div} \boldsymbol{E} = \frac{\varrho}{\varepsilon_0},$$

$$\mathbf{curl}\,\boldsymbol{H} = \varepsilon_0 \frac{\partial \boldsymbol{E}}{\partial t} + \boldsymbol{J}, \qquad \operatorname{div} \boldsymbol{H} = 0.$$

Thus

$$\begin{aligned}\mathbf{curl\ curl}\,\boldsymbol{E} &= \mathbf{grad}\,(\varrho/\varepsilon_0) - \nabla^2 \boldsymbol{E}\\ &= -\mu_0 \frac{\partial}{\partial t}\left\{\varepsilon_0 \frac{\partial \boldsymbol{E}}{\partial t} + \boldsymbol{J}\right\},\end{aligned}$$

so that

$$\frac{1}{c^2}\frac{\partial^2 \boldsymbol{E}}{\partial t^2} - \nabla^2 \boldsymbol{E} = -\mu_0 \frac{\partial \boldsymbol{J}}{\partial t} - \frac{1}{\varepsilon_0}\nabla \varrho. \tag{13.32}$$

Similarly

$$\frac{1}{c^2}\frac{\partial^2 \boldsymbol{H}}{\partial t^2} - \nabla^2 \boldsymbol{H} = \mathbf{curl}\,\boldsymbol{J}. \tag{13.33}$$

It is to be noticed that the sources on the right-hand sides of these equations are very intractable, because they involve the derivatives of ϱ and $\boldsymbol{J}$. This suggests a simpler situation, where we can write $\boldsymbol{E}, \boldsymbol{H}$ as derivatives; in other words, if one introduces the potentials.

When we write, then, as before

$$\boldsymbol{H} = \frac{1}{\mu_0}\,\mathbf{curl}\,\boldsymbol{A},$$

$$\boldsymbol{E} = -\nabla\phi - \frac{\partial \boldsymbol{A}}{\partial t},$$

it follows that

$$-\nabla^2\phi - \frac{\partial}{\partial t}(\text{div } \boldsymbol{A}) = \varrho/\varepsilon_0$$

and

$$\textbf{curl curl } \boldsymbol{A} = -\frac{1}{c^2}\frac{\partial^2 \boldsymbol{A}}{\partial t^2} - \nabla\left(\frac{1}{c^2}\frac{\partial \phi}{\partial t}\right) + \mu_0 \boldsymbol{J}.$$

But

$$\textbf{curl curl } \boldsymbol{A} = \textbf{grad} \text{ div } \boldsymbol{A} - \nabla^2 \boldsymbol{A},$$

so that

$$\frac{1}{c^2}\frac{\partial^2 \boldsymbol{A}}{\partial t^2} - \nabla^2 \boldsymbol{A} = \mu_0 \boldsymbol{J} + \nabla\left(\text{div } \boldsymbol{A} + \frac{1}{c^2}\frac{\partial \phi}{\partial t}\right).$$

If we again fix $\boldsymbol{A}$, ϕ by the Lorentz condition

$$\text{div } \boldsymbol{A} + \frac{1}{c^2}\frac{\partial \phi}{\partial t} = 0$$

it follows that $\boldsymbol{A}$ will satisfy the (inhomogeneous) wave equation. Moreover, ϕ will satisfy

$$\frac{1}{c^2}\frac{\partial^2 \phi}{\partial t^2} - \nabla^2 \phi = \frac{\varrho}{\varepsilon_0}.$$

A word is appropriate here about the solution of these wave equations. A common method in some theories (e.g. acoustics), where the scalar wave equation arises, is to seek for coordinate systems in which there exist separable solutions. In point of fact, there are known to exist exactly eleven such coordinate systems, of which cartesian, spherical polars and cylindrical polars are the best known. But here we have to solve a vector wave equation also; in every coordinate system but cartesians the operator ∇^2 must be interpreted as $-$**curl curl**$+$**grad** div. As a result one component of the operation of ∇^2 on a vector involves the other components also. Very little is known about the separable solutions of this equation. Accordingly we must have recourse to other methods.

Consider then the scalar wave equation above. As usual with differential equations of the inhomogeneous type, the complete solution consists of any solution of the homogeneous equation

$$\frac{1}{c^2}\frac{\partial^2 \phi}{\partial t^2} - \nabla^2 \phi = 0$$

(complementary function), together with some particular integral of the original equation. To find such a particular integral let us divide the space up into very small portions and find the separate fields produced by the charges in each of these small elements. The resultant of all these fields will then be a particular integral. For any given volume element $d\tau$ containing a charge $\varrho\, d\tau$ we have then to solve the homogeneous wave equation everywhere, except in the immediate neighbourhood of the volume element where the charge corresponds to a point charge $\varrho\, d\tau$ (which is a function of time). Choosing, for a moment, the origin at the volume element, we have to solve the wave equation which will be (for a spherically symmetric partial function $\delta\phi$)

$$\frac{1}{c^2}\frac{\partial^2(\delta\phi)}{\partial t^2}-\frac{1}{r^2}\frac{\partial}{\partial r}\left(r^2\frac{\partial(\delta\phi)}{\partial r}\right)=0.$$

As usual, we can solve this by writing $\delta\phi = ru$, when we find that

$$\frac{1}{c^2}\frac{\partial^2 u}{\partial t^2}-\frac{\partial^2 u}{\partial r^2}=0, \tag{13.34}$$

which has the solution $u = f(t-r/c)+g(t+r/c)$, for any functions f, g. We are here looking for *one* particular solution, so we may choose one of the functions f, g to be zero. In most elementary cases, the most convenient choice is $g = 0$. This corresponds to a solution in which a disturbance in the charge produces effects in the field afterwards rather than before it happens. But it by no means follows from this rough appeal to causality that the same choice is always required. Certainly Maxwell's theory allows the choice $f = 0$, which has been put forward in some theories of the classical electron. However, for our present purposes it will be sufficient to confine ourselves to the case $g = 0$.

As a result $\delta\phi = (1/r)\, f(t-r/c)$; and the function f is still at our disposal. We have to choose it so as to make the potential correspond to the correct value of the charge, that is, to make the potential have the correct value very near the origin, as $r \to 0$. Accordingly, for our case

$$f(t) = \varrho(t)\frac{d\tau}{4\pi\varepsilon_0}.$$

We must now transform to a general origin, and so derive

$$\delta\phi(\boldsymbol{r}) = \frac{\varrho(t-r/c,\, \boldsymbol{r}')\, d\tau'}{4\pi\varepsilon_0 r}$$

where $r = |\boldsymbol{r}-\boldsymbol{r}'|$, and $\boldsymbol{r}'$ is the point occupied by charge (so that $d\tau'$ is the volume element at that point) whilst $\boldsymbol{r}$ is the position of the field point, at which $\delta\phi$ is required. The total field is therefore

$$\phi(\boldsymbol{r}) = \int \frac{\varrho(t-r/c, \boldsymbol{r}')\,d\tau'}{4\pi\varepsilon_0 r}$$

$$= \int \frac{[\varrho]\,d\tau'}{4\pi\varepsilon_0 r} \tag{13.35}$$

where we introduce square brackets to mean the so-called *retarded value*. To this field we should in general add any solution of the homogeneous wave equation which is then chosen so as to satisfy the conditions of the problem. For our purposes, however, it will be sufficient merely to consider in detail the particular integral found (called the *retarded solution*). By writing the vector wave equation in cartesians, a similar argument on each component leads to the result

$$\boldsymbol{A}(\boldsymbol{r}) = \mu_0 \int \frac{[\boldsymbol{J}]\,d\tau'}{4\pi r}. \tag{13.36}$$

13.8 The Liénard–Wiechert potentials

The general formulae which we have just found for ϕ, $\boldsymbol{A}$ are not usually very easy to apply in practice. But their chief importance is that they express the potentials of the field of a moving charge in an explicit form which, as we can see from the expression, depends on its velocity but not on its acceleration.

Suppose that we have a single charge, moving in any manner. We have to consider the limiting case of these formulae as the density corresponds to the singular one of the point charge. However, a little care is needed in the limiting procedure. We cannot simply replace $[\varrho]\,d\tau$ by e since we have initially a distribution of charge, and different parts of it will have different retarded times. It is not true that it makes no difference whether we carry out the limiting operation before or after taking retarded values. There is only one frame of reference in which this is true—that in which the charge is instantaneously at rest. In this frame of reference the potentials are

$$\phi = \frac{e}{4\pi\varepsilon_0 r}, \quad \boldsymbol{A} = \boldsymbol{0}. \tag{13.37}$$

Now we use two facts: first, from the form of the expression for ϕ, $\boldsymbol{A}$ in a general reference frame, it is clear that they will depend only on the velocity of the charge and not on its acceleration. Second, we are concerned with retarded values; so that the signals at a point (say, the origin) at time t, from a moving charge e at $\boldsymbol{r} = \boldsymbol{r}(t)$ will arise from the charge when it is at a point $\boldsymbol{r}(t')$, where

$$|\boldsymbol{r}(t')| = c(t-t').$$

Let us first apply this formula in the frame of reference in which the charge is at rest. Then

$$\phi = \frac{e}{4\pi\varepsilon_0 c(t-t')}, \quad \boldsymbol{A} = \boldsymbol{0}$$

in this frame of reference.

In order to see the value in any other frame of reference it is necessary to express everything in four-vector form. The velocity four-vector, for the charge, is defined by $(\beta, \beta\boldsymbol{v})$ where β is, as usual, $(1-v^2/c^2)^{-1/2}$. Accordingly in the rest-frame of the charge, it is $(1, \boldsymbol{0})$. Moreover, the combination $(\phi/c^2, \boldsymbol{A})$ is also a four-vector, and has in this reference frame the value

$$\left(\frac{\mu_0 e}{4\pi c(t-t')}, \boldsymbol{0}\right). \tag{13.38}$$

These two four-vectors are proportional, and we should therefore expect the factor of proportionality to be independent of the reference frame. Now the condition for the retarded time, above, can be written

$$c^2(t-t')^2 - \boldsymbol{r}^2 = 0,$$

which is certainly an invariant relation for the four-vector $(t-t', -\boldsymbol{r})$ obtained by subtracting the two four-vectors $(t, \boldsymbol{0})$ and $(t', \boldsymbol{r}(t')) = (t', \boldsymbol{r}')$. Now from this four-vector and the velocity, in the reference frame in which the particle is at rest, the invariant

$$s = c^2\beta(t-t') + \beta\boldsymbol{v}\cdot\boldsymbol{r} \tag{13.39}$$

has the value $c^2(t-t')$. Hence in this frame of reference the potential four-vector is derived from the velocity four-vector by multiplying it by $\mu_0 ce/(4\pi s)$. Since this factor is invariant the same is true in every reference frame, so that, in general

$$\frac{\phi}{c^2} = \frac{\mu_0 ce\beta}{4\pi s}, \quad \boldsymbol{A} = \frac{\mu_0 ce\beta\boldsymbol{v}}{4\pi s}. \tag{13.40}$$

It only remains to put these into three-dimensional form. For this purpose it is more useful to go over to having the origin chosen at the charge, and to consider the field at a point $\boldsymbol{r}$; this corresponds to a change of sign of $\boldsymbol{r}$ in these formulae. Using

$$c^2(t-t')^2 = \boldsymbol{r}^2,$$

we have, with the understanding that everything on the right-hand side is evaluated at the retarded time t',

$$\frac{\phi}{c^2} = \frac{\mu_0 e}{4\pi\left(r-\frac{\boldsymbol{r}\cdot\boldsymbol{v}}{c}\right)} \quad \text{or} \quad \phi = \frac{e}{4\pi\varepsilon_0\left(r-\frac{\boldsymbol{r}\cdot\boldsymbol{v}}{c}\right)}, \tag{13.41}$$

$$\text{and} \quad \boldsymbol{A} = \frac{\mu_0 e\boldsymbol{v}}{4\pi\left(r-\frac{\boldsymbol{r}\cdot\boldsymbol{v}}{c}\right)}, \quad \text{where} \quad r = |\boldsymbol{r}|. \tag{13.42}$$

These are the Liénard–Wiechert potentials.

13.9 Calculation of the field strengths

We now have to use the formulae

$$\boldsymbol{E} = -\nabla\phi - \frac{\partial \boldsymbol{A}}{\partial t}, \quad \boldsymbol{H} = \frac{1}{\mu_0}\,\textbf{curl}\,\boldsymbol{A}$$

to determine the field strengths. But the differentiations here are all with respect to t and $\boldsymbol{r}$, whereas the Liénard–Wiechert potentials are given in terms of t'. Some care is therefore needed in the differentiation.

Since $r(t') = c(t-t')$, it follows that

$$\frac{\partial r}{\partial t} = c\left(1-\frac{\partial t'}{\partial t}\right). \tag{13.43}$$

But we can find an alternative expression by differentiating in the form

$$\frac{\partial r}{\partial t} = \frac{\partial t'}{\partial t}\frac{\partial r}{\partial t'}.$$

Now

$$r^2 = \boldsymbol{r}^2,$$

so that

$$r\frac{\partial r}{\partial t'} = \boldsymbol{r}\cdot\frac{\partial \boldsymbol{r}}{\partial t'}$$

and $\partial\boldsymbol{r}/\partial t' = -\boldsymbol{v}$ (by definition). Hence

$$\frac{\partial r}{\partial t'} = -\frac{\boldsymbol{r}\cdot\boldsymbol{v}}{r}$$

and

$$\frac{\partial r}{\partial t} = -\frac{\boldsymbol{r}\cdot\boldsymbol{v}}{r}\frac{\partial t'}{\partial t}.$$

Collecting these results together, it follows that

$$\frac{\partial t'}{\partial t} = \frac{1}{1-\dfrac{\boldsymbol{r}\cdot\boldsymbol{v}}{rc}}. \tag{13.44}$$

Using $r(t') = c(t-t')$, it also follows that

$$\nabla t' = -\frac{1}{c}\nabla r,$$

$$\nabla t' = -\frac{1}{c}\left\{\frac{\partial r}{\partial t'}\nabla t' + \frac{\boldsymbol{r}}{r}\right\}.$$

Hence
$$\nabla t' = -\frac{\boldsymbol{r}}{c\left(r-\dfrac{\boldsymbol{r}\cdot\boldsymbol{v}}{c}\right)}. \tag{13.45}$$

In carrying out the calculations, it is convenient to use two subsidiary definitions:

$$R = r - \frac{\boldsymbol{r}\cdot\boldsymbol{v}}{c}, \tag{13.46}$$

and
$$\boldsymbol{S} = \boldsymbol{r} - \frac{r\boldsymbol{v}}{c}, \tag{13.47}$$

noticing that, when both *t and* t' are held constant

$$\nabla R = \hat{\boldsymbol{r}} - \frac{\boldsymbol{v}}{c} = \frac{\boldsymbol{S}}{r},$$

and
$$\boldsymbol{r}\cdot\boldsymbol{S} = rR.$$

Moreover, at a fixed point,

$$\frac{\partial R}{\partial t'} = \dot{r} - \frac{\dot{\boldsymbol{r}}\cdot\boldsymbol{v}}{c} - \frac{\boldsymbol{r}\cdot\dot{\boldsymbol{v}}}{c} = \frac{\boldsymbol{r}\cdot\dot{\boldsymbol{r}}}{r} - \frac{\dot{\boldsymbol{r}}\cdot\boldsymbol{v}}{c} - \frac{\boldsymbol{r}\cdot\dot{\boldsymbol{v}}}{c}$$

and $\dot{\boldsymbol{r}} = -\boldsymbol{v}$, so that

$$\frac{\partial R}{\partial t'} = -\frac{\boldsymbol{r}\cdot\boldsymbol{v}}{r} + \frac{v^2}{c} - \frac{\boldsymbol{r}\cdot\dot{\boldsymbol{v}}}{c} = -\frac{\boldsymbol{v}\cdot\boldsymbol{S}}{r} - \frac{\boldsymbol{r}\cdot\dot{\boldsymbol{v}}}{c}.$$

The potentials now take the form

$$\phi = \frac{e}{4\pi\varepsilon_0 R}, \quad \boldsymbol{A} = \frac{\mu_0 e\boldsymbol{v}}{4\pi R}, \tag{13.48}$$

so that

$$4\pi \boldsymbol{E} = -\nabla\left(\frac{e}{\varepsilon_0 R}\right) - \frac{\partial}{\partial t}\left(\frac{\mu_0 e \boldsymbol{v}}{R}\right).$$

Notice here that the gradient is to be evaluated when t is held constant, not t', so that, collecting all the results together,

$$\boldsymbol{E} = \frac{e\boldsymbol{S}}{4\pi\varepsilon_0 R^2 r} + \frac{e}{4\pi\varepsilon_0 R^2}\left(\frac{\boldsymbol{v}\cdot\boldsymbol{S}}{r} + \frac{\boldsymbol{r}\cdot\dot{\boldsymbol{v}}}{c}\right)\frac{\boldsymbol{r}}{cR}$$

$$- \frac{\mu_0 e\dot{\boldsymbol{v}}}{4\pi R}\frac{r}{R} - \frac{\mu_0 e\boldsymbol{v}}{4\pi R^2}\left(\frac{\boldsymbol{v}\cdot\boldsymbol{S}}{r} + \frac{\boldsymbol{r}\cdot\boldsymbol{v}}{c}\right)\frac{r}{R}. \tag{13.49}$$

The details do not matter so much as the fact that $\boldsymbol{E}$ contains some terms that are independent of the acceleration, and these terms fall off, roughly, as the square of the distance (counting, for this purpose, $\boldsymbol{r}$, R and $\boldsymbol{S}$ as all of the same size), whilst there are other terms proportional to the acceleration and these—the characteristic radiation terms—fall off as the distance.

In fact the terms independent of the acceleration become (remembering that $c^2\mu_0\varepsilon_0 = 1$)

$$\boldsymbol{E} = \frac{e}{4\pi\varepsilon_0 R^2}\left\{\frac{\boldsymbol{S}}{r} + \frac{(\boldsymbol{v}\cdot\boldsymbol{S})\boldsymbol{r}}{cRr} - \frac{1}{c^2}\frac{\boldsymbol{v}(\boldsymbol{v}\cdot\boldsymbol{S})}{R}\right\} = \frac{e}{4\pi\varepsilon_0 R^2}\frac{1}{r}\left(1 + \frac{\boldsymbol{v}\cdot\boldsymbol{S}}{cR}\right)\boldsymbol{S}. \tag{13.50}$$

(If $\boldsymbol{v} = \boldsymbol{0}$, so that the charge is at rest, this at once reduces to the electrostatic form.) For the radiation terms, we have

$$\boldsymbol{E} = \frac{e}{4\pi\varepsilon_0 R^2}\left[\frac{\boldsymbol{r}\cdot\dot{\boldsymbol{v}}\boldsymbol{r}}{c^2 R} - \frac{\dot{\boldsymbol{v}}r}{c^2} - \frac{\boldsymbol{v}r\boldsymbol{r}\cdot\dot{\boldsymbol{v}}}{c^3 R}\right]$$

$$= \frac{e}{4\pi\varepsilon_0 R^2}\left[\frac{\boldsymbol{r}\cdot\dot{\boldsymbol{v}}}{c^2 R}\left(\boldsymbol{r} - \frac{r\boldsymbol{v}}{c}\right) - \frac{\dot{\boldsymbol{v}}r}{c^2}\right]$$

$$= \frac{e}{4\pi\varepsilon_0 c^2 R^3}[(\boldsymbol{r}\cdot\dot{\boldsymbol{v}})\boldsymbol{S} - Rr\dot{\boldsymbol{v}}]. \tag{13.51}$$

But we noticed before that

$$Rr = \boldsymbol{r}\cdot\boldsymbol{S},$$

so that the expression for $\boldsymbol{E}$ can be rewritten

$$\boldsymbol{E} = \frac{\mu_0 e}{4\pi R^3}[\boldsymbol{r}\times(\boldsymbol{S}\times\dot{\boldsymbol{v}})]. \tag{13.52}$$

The first field (independent of the acceleration) is evidently the field of charge in uniform motion and this expression could have been derived by making a Lorentz transformation of the field strengths directly. The second, however, depends on the acceleration, and so is non-zero even for a charge which is instantaneously at rest. In fact, for such a charge, with acceleration $\boldsymbol{f}$, the total field consists of the Coulomb field together with a radiation component of

$$\boldsymbol{E} = \frac{\mu_0 e}{4\pi r} \hat{\boldsymbol{r}} \times (\hat{\boldsymbol{r}} \times \boldsymbol{f}). \tag{13.53}$$

The calculation of the magnetic field is carried out in the same way, and it turns out that

$$\begin{aligned} \boldsymbol{H} &= \sqrt{\left(\frac{\varepsilon_0}{\mu_0}\right)} \frac{1}{r} \boldsymbol{r} \times \boldsymbol{E} \\ &= \frac{1}{cr} \boldsymbol{r} \times \left(\frac{\boldsymbol{E}}{\mu_0}\right). \end{aligned} \tag{13.54}$$

Thus the magnetic field is everywhere perpendicular to the electric field. This fact was to be expected of the radiation parts of the fields but it is surprising that it holds of the total field.

It remains to draw the readers' attention to one disquieting fact about these results. We began by solving the wave equation for the potentials, found a solution that depended only on velocity and not on acceleration, and then differentiated this solution in order to derive a field that depended both on velocity and on acceleration. This suggests that, if we had begun with the Hertz potential being independent of the acceleration, then the ordinary potentials would have depended on acceleration, and the fields on its derivative. Does our argument forbid this? Some discussion has taken place on this subject, and the matter is still not clear. It is our opinion, however, that the argument we have given contains a hidden assumption. When we solved the scalar wave equation

$$\frac{1}{c^2} \frac{\partial^2 \phi}{\partial t^2} - \nabla^2 \phi = \frac{\varrho}{\varepsilon_0},$$

we sought a particular integral that was spherically symmetric, about an origin chosen at the position of charge involved. If ϕ were to depend on the acceleration—which would then have a definite direction, even in the rest-frame of the charge—this assumption of spherical symmetry would no longer be justified. Instead we would need axially symmetric solutions, of which, of course, there are many, involving Legendre polynomials. Conversely, when we make the assumption of spherical symmetry, we are assuming that the accelerations are unimportant here. The solution that we have derived must then rely, for its complete justification, on its experimental confirmation, and fortunately there is ample experimental verification that it is, in fact, true.

Example 1. An infinite uniform wire occupies the z-axis in free space, the current in it vanishing for $t \leqslant 0$ and having the constant value J for $t > 0$. Show that the vector potential at a point P, whose cylindrical coordinates are r, ϕ, z, is given by

$$\mathbf{A} = \mu_0 \frac{J}{2\pi} \hat{\mathbf{k}} \ln \frac{ct+(c^2t^2-r^2)^{1/2}}{r} \quad \text{for} \quad t \geqslant r/c,$$

and

$$\mathbf{A} = \mathbf{0} \quad \text{for} \quad t < r/c,$$

and that the electromagnetic field of the current at P when $t \geqslant r/c$ is given by

$$\mathbf{E} = \frac{-J\hat{\mathbf{k}}c}{2\pi(c^2t^2-r^2)^{1/2}}, \quad \mu_0 \mathbf{H} = \frac{Jct\hat{\boldsymbol{\varphi}}}{2\pi r(c^2t^2-r^2)^{1/2}}.$$

Find the flux of energy across unit length of a cylinder passing through P and coaxial with the wire, when $t \geqslant r/c$.

(i) Using formula (13.36) in the text

$$\mathbf{A}(\mathbf{r}, t) = \frac{\mu_0}{4\pi} \int \frac{[J\mathbf{k}]\,\mathrm{d}z_1}{\sqrt{\{(z_1-z)^2+x^2+y^2\}}} = \frac{1}{4\pi} \int \frac{[J\mathbf{k}]\,\mathrm{d}z_1}{\sqrt{\{(z_1-z)^2+r^2\}}},$$

where the part of the wire involved is such that

$$t - \frac{\sqrt{\{(z_1-z)^2+r^2\}}}{c} \geqslant 0,$$

i.e.

$$(z_1-z)^2 \leqslant c^2t^2-r^2$$

so that z_1-z ranges from $\pm\sqrt{(c^2t^2-r^2)}$, if $ct \geqslant r$.

Changing variables gives

$$\mathbf{A}(\mathbf{r}, t) = \frac{\mu_0 J\mathbf{k}}{4\pi} \int_{-\sqrt{(c^2t^2-r^2)}}^{\sqrt{(c^2t^2-r^2)}} \frac{\mathrm{d}\lambda}{\sqrt{(\lambda^2+r^2)}} = \frac{\mu_0 J\mathbf{k}}{2\pi} \left[\sinh^{-1}\left(\frac{\lambda}{r}\right)\right]_0^{\sqrt{(c^2t^2-r^2)}}$$

$$= \frac{\mu_0 J\mathbf{k}}{2\pi} \sinh^{-1}\left\{\frac{\sqrt{(c^2t^2-r^2)}}{r}\right\} = \frac{\mu_0 J\mathbf{k}}{2\pi} \cosh^{-1}\left(\frac{ct}{r}\right)$$

$$= \frac{\mu_0 J\mathbf{k}}{2\pi} \ln\left\{\frac{ct+\sqrt{(c^2t^2-r^2)}}{r}\right\}, \qquad (ct \geqslant r).$$

If $ct \leqslant r$, the field is obviously zero.

(ii) For the field vectors

$$\mathbf{E} = -\frac{\partial \mathbf{A}}{\partial t} = -\frac{\partial}{\partial t}\left\{\frac{\mu_0 J\mathbf{k}}{2\pi} \cosh^{-1}\left(\frac{ct}{r}\right)\right\}$$

$$= -\frac{\mu_0 J\mathbf{k}}{2\pi} \frac{c/r}{\sqrt{[(c^2t^2/r^2)-1]}} = \frac{-\mu_0 J\mathbf{k}c}{2\pi\sqrt{(c^2t^2-r^2)}}.$$

Similarly

$$\mu_0 \boldsymbol{H} = \mathbf{curl}\left\{\frac{\mu_0 J\boldsymbol{k}}{2\pi}\cosh^{-1}\left(\frac{ct}{r}\right)\right\},$$

$$\boldsymbol{H} = \frac{J}{2\pi}\hat{\boldsymbol{r}}\times\boldsymbol{k}\frac{\partial}{\partial r}\left\{\cosh^{-1}\left(\frac{ct}{r}\right)\right\}$$

$$= -\frac{J}{2\pi}\hat{\boldsymbol{\phi}}\frac{(-ct/r^2)}{\sqrt{[(c^2t^2/r^2)-1]}}$$

$$= \frac{Jct\hat{\boldsymbol{\phi}}}{2\pi r\sqrt{(c^2t^2-r^2)}}.$$

(iii) Finally $\boldsymbol{E}\times\boldsymbol{H} = \dfrac{\mu_0 J^2c^2t\hat{\boldsymbol{r}}}{4\pi^2 r(c^2t^2-r^2)}$ which then gives the flux of energy.

Example 2. Show that in a vacuum Maxwell's equations are satisfied by

$$E_\theta = -\frac{1}{r}\frac{\partial^2 u}{\partial\theta\,\partial r},\quad E_\phi = -\frac{1}{r\sin\theta}\frac{\partial^2 u}{\partial\phi\,\partial r},\quad E_r = -\left(\frac{\partial^2 u}{\partial r^2}-\frac{1}{c^2}\frac{\partial^2 u}{\partial t^2}\right),$$

$$H_\theta = -\frac{\varepsilon_0}{r\sin\theta}\frac{\partial^2 u}{\partial\phi\,\partial t},\quad H_\phi = \frac{\varepsilon_0}{r}\frac{\partial^2 u}{\partial\theta\,\partial t},\quad H_r = 0,$$

where r, θ, ϕ are spherical polar coordinates and u satisfies a certain differential equation. Show that if the field is symmetrical about $\theta = 0$ this equation has a solution

$$u = P_n(\cos\theta)\, r^{n+1}\left(\frac{1}{r}\frac{\partial}{\partial r}\right)^n\left\{\frac{1}{r}f\left(t-\frac{r}{c}\right)\right\}.$$

From the double differentiations involved in the expressions for the field quantities, it is clear that some analogue of the Hertz potential is required. Recalling that

$$\boldsymbol{H} = \varepsilon_0\,\mathbf{curl}\,\dot{\boldsymbol{\Pi}}$$

suggests that we try a vector $\boldsymbol{\Pi}$ whose components of **curl** in spherical polars are

$$\left\{0\quad -\frac{1}{r\sin\theta}\frac{\partial u}{\partial\phi}\quad \frac{1}{r}\frac{\partial u}{\partial\theta}\right\}.$$

But a comparison with the general expression for **curl** in spherical polars shows that $\boldsymbol{\Pi}$ must then be radial, of magnitude $-u$. With this choice

$$\boldsymbol{H} = \varepsilon_0\,\mathbf{curl}\,\dot{\boldsymbol{\Pi}}$$

so that div $\boldsymbol{H} = 0$. Next calculate

$$\mathbf{curl}\,\mathbf{curl}\,\boldsymbol{\Pi} = \left\{\frac{1}{r^2\sin\theta}\left[\frac{\partial}{\partial\theta}\left(\sin\theta\frac{\partial u}{\partial\theta}\right)\right]+\frac{1}{r^2\sin^2\theta}\frac{\partial^2 u}{\partial\phi^2}\quad -\frac{1}{r}\frac{\partial^2 u}{\partial r\,\partial\theta}\quad -\frac{1}{r\sin\theta}\frac{\partial^2 u}{\partial r\,\partial\phi}\right\}.$$

This will agree with the given solution only if u satisfies the differential equation

$$\frac{1}{c^2}\frac{\partial^2 u}{\partial t^2} = \frac{\partial^2 u}{\partial r^2}+\frac{1}{r^2\sin\theta}\frac{\partial}{\partial\theta}\left(\sin\theta\frac{\partial u}{\partial\theta}\right)+\frac{1}{r^2\sin^2\theta}\frac{\partial^2 u}{\partial\phi^2}.$$

(Note that this is *not* the wave equation in spherical polars!) If u satisfies this equation, then we have

$$\boldsymbol{E} = \mathbf{curl\ curl}\ \boldsymbol{\Pi}$$

so that

$$\text{div}\ \boldsymbol{E} = 0$$

and

$$\mathbf{curl}\ \boldsymbol{H} = \varepsilon_0 \frac{\partial \boldsymbol{E}}{\partial t} = \frac{\partial \boldsymbol{D}}{\partial t}.$$

Three of Maxwell's equations are therefore satisfied, and it remains only to verify that $\mathbf{curl}\ \boldsymbol{E} = -\mu_0(\partial \boldsymbol{H}/\partial t)$.

The first step is to calculate the r-component of $\mathbf{curl}\ \boldsymbol{E}$ thus:

$$-\frac{1}{r \sin\theta}\left\{\frac{\partial}{\partial\theta}\left(\sin\theta \frac{1}{r\sin\theta}\frac{\partial^2 u}{\partial\phi\,\partial r}\right) - \frac{\partial}{\partial\phi}\left(\frac{1}{r}\frac{\partial^2 u}{\partial\theta\,\partial r}\right)\right\} = 0.$$

In calculating the other components the terms involving mixed derivatives with respect to r, θ, ϕ cancel, leaving

$$\mathbf{curl}\ \boldsymbol{E} = \left\{0 \quad \frac{1}{c^2 r \sin\theta}\frac{\partial^3 u}{\partial\phi\,\partial t^2} \quad -\frac{1}{c^2 r}\frac{\partial^3 u}{\partial\theta\,\partial t^2}\right\}$$

$$= -\varepsilon_0\mu_0 \frac{\partial(\boldsymbol{H}/\varepsilon_0)}{\partial t}$$

as required.

If the solution is independent of ϕ, one term vanishes in the "wave" equation. By exactly the same treatment, as in the case of Laplace's equation,

$$\frac{1}{\sin\theta}\frac{\partial}{\partial\theta}\left(\sin\theta\frac{\partial f}{\partial\theta}\right) = -n(n+1)f$$

if $f = P_n(\cos\theta)$, so that a separable solution is given by $u = A(r, t)\,P_n(\cos\theta)$, where

$$\frac{1}{c^2}\frac{\partial^2 A}{\partial t^2} - \frac{\partial^2 A}{\partial r^2} + \frac{n(n+1)}{r^2}A = 0.$$

The given solution in the case $n = 0$ is obvious.

Consider next the case $n = 1$. Then the given solution is

$$A = r\frac{\partial C}{\partial r},$$

where C satisfies

$$\frac{1}{c^2}\frac{\partial^2 C}{\partial t^2} = \frac{\partial^2 C}{\partial r^2} + \frac{2}{r}\frac{\partial C}{\partial r},$$

i.e. the true wave equation. In fact,

$$\frac{1}{c^2}\frac{\partial^2 A}{\partial t^2} = r\frac{\partial}{\partial r}\left(\frac{1}{c^2}\frac{\partial^2 C}{\partial t^2}\right)$$

$$= r\frac{\partial}{\partial r}\left(\frac{\partial^2 C}{\partial r^2} + \frac{2}{r}\frac{\partial C}{\partial r}\right)$$

$$= r\frac{\partial^3 C}{\partial r^3} + 2\frac{\partial^2 C}{\partial r^2} - \frac{2}{r}\frac{\partial C}{\partial r}$$

$$= \frac{\partial^2 A}{\partial r^2} - \frac{2}{r^2}A$$

which is the correct equation.

In the general case, if

$$u_n = r^{n+1}\left(\frac{1}{r}\frac{\partial}{\partial r}\right)^n C$$

and

$$u_{n+1} = r^{n+2}\left(\frac{1}{r}\frac{\partial}{\partial r}\right)^{n+1} C,$$

then

$$u_{n+1} = r^{n+1}\frac{\partial}{\partial r}\left(\frac{u_n}{r^{n+1}}\right)$$

and a similar piece of working proves the result by induction.

Example 3. Show that a charge moving in a circle with constant speed u emits no radiation in the plane of the circle in directions inclined at an angle $\cos^{-1}(u/c)$ to the direction of motion.

The radiation fields contain the factors

$$\boldsymbol{E} \propto \boldsymbol{r}\times\left\{\left(\boldsymbol{r}-\frac{r\boldsymbol{v}}{c}\right)\times\dot{\boldsymbol{v}}\right\},$$

$$\boldsymbol{B} = \frac{1}{c}\hat{\boldsymbol{r}}\times\boldsymbol{E}.$$

Hence

$$\boldsymbol{E}\times\boldsymbol{H} \propto \boldsymbol{E}\times(\hat{\boldsymbol{r}}\times\boldsymbol{E}) = E^2\hat{\boldsymbol{r}}$$

since $\boldsymbol{E}\cdot\hat{\boldsymbol{r}} \equiv 0$. Thus the direction in which there is no flux of radiation is the direction in which $\boldsymbol{E}$ vanishes. For the circular motion there is one obvious way of satisfying this, by making

$$\left(\boldsymbol{r}-\frac{r\boldsymbol{v}}{c}\right) \text{ parallel to } \dot{\boldsymbol{v}}.$$

Since $\dot{\boldsymbol{v}}$ is along the radius to the circle this shows at once, from Fig. 13.2, that $\sin\alpha = u/c$, which gives the result.

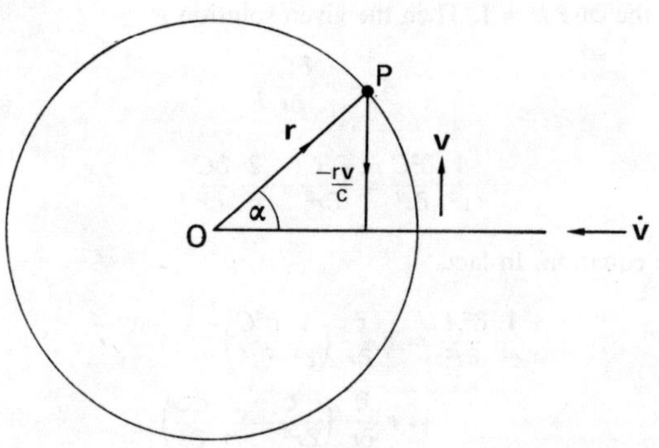

FIG. 13.2

But might there be other directions for which $\boldsymbol{E} = \boldsymbol{0}$? It is clear that there cannot be, since $\boldsymbol{r}$, $\boldsymbol{v}$ and $\dot{\boldsymbol{v}}$ are all in the plane of the circle, so that the contents of the braces are normal to this plane and cannot be parallel to a direction $\boldsymbol{r}$ in the plane.

Example 4. Two Hertzian radiators P_1, P_2 are oscillating in phase in free space. Each has dipole moment $p_0 e^{i\omega t}\boldsymbol{k}$, and the position vector of P_2 relative to P_1 is $a\boldsymbol{k}$, $\boldsymbol{k}$ being a unit vector. Show that the electromagnetic field at the point $\boldsymbol{r}$ relative to P_1 is the field of P_1 alone multiplied by the factor

$$1+\exp\{i\omega a\boldsymbol{k}\cdot\boldsymbol{r}/(cr)\},$$

provided $a \ll r$.

If $a = n\lambda$, where λ is the wavelength and n an integer, discuss the radiation pattern in the wave zone. Show that the ratio of the total energy radiated to that radiated by two non-interfering oscillators is

$$1-\frac{3}{4(n\pi)^2}.$$

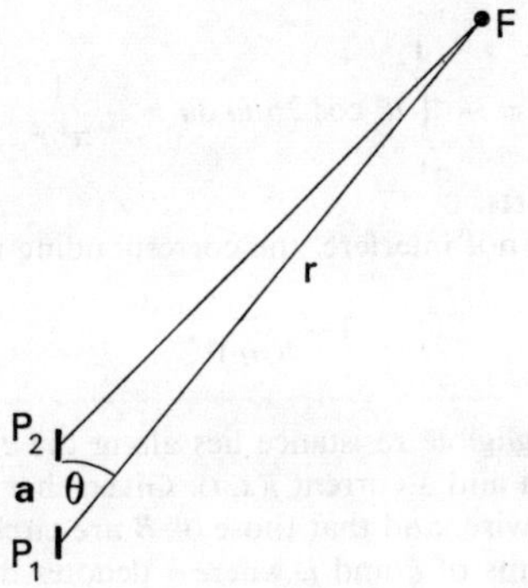

FIG. 13.3

With reference to Fig. 13.3 it is a sufficient approximation to consider the field point F equidistant from the two oscillators as far as the field *strengths* are concerned, but it is necessary to take account of the difference in calculating the retarded time. Then

$$P_2F = (r^2+a^2-2ar\cos\theta)^{1/2} \approx r\left(1-\frac{2a}{r}\cos\theta\right)^{1/2} \approx r-a\cos\theta.$$

Since, from the text, E_θ and H_ϕ are each proportional to $\ddot{p} = -\omega^2 p e^{i\omega t}$, the total field is proportional to

$$\exp\left\{i\omega\left(t-\frac{r}{c}\right)\right\}+\exp\left\{i\omega\left(t-\frac{r}{c}+\frac{a\cos\theta}{c}\right)\right\}$$

$$= \exp\left\{i\omega\left(t-\frac{r}{c}\right)\right\}\left[1+\exp\left\{\frac{(i\omega a\cos\theta)}{c}\right\}\right],$$

which is the result given, since $\cos\theta = \boldsymbol{r}\cdot\boldsymbol{k}/r$.

If $a = n\lambda$, then, since $\lambda\nu = c$ and $\omega = 2\pi\nu$, it follows that $a = 2\pi nc/\omega$, so that the factor is $1+e^{2\pi i n\cos\theta}$. The energy radiated at an angle θ is proportional to the square of the modulus [since $(\alpha+i\beta)e^{i\mu} = |\alpha+i\beta|e^{i(\mu+\gamma)}$ where $\gamma = \tan^{-1}\beta/\alpha$], i.e. to

$$[1+\exp\{(i\omega a\cos\theta)/c\}]\,[1+\exp\{(-i\omega a\cos\theta)/c\}] = 2+2\cos\{(\omega a\cos\theta)/c\}$$

$$= 2[1+\cos(2\pi n\cos\theta)].$$

There are therefore nodal lines whenever $\cos(2\pi n\cos\theta) = -1$,

i.e. $$\cos\theta = \frac{2k+1}{2n} \qquad (k = 0, \ldots, n-1).$$

The energy emitted at an angle θ is proportional to $[1+\cos(2\pi n\cos\theta)]\sin^2\theta$, since the factor $\sin\theta$ occurs in the expressions for the field strengths. The total energy is therefore proportional to

$$\int_0^\pi [1+\cos(2\pi n\cos\theta)]\sin^2\theta\sin\theta\, d\theta$$

$$= +\tfrac{4}{3}+\int_0^\pi \cos(2\pi n\cos\theta)\sin^2\theta\sin\theta\, d\theta.$$

The second integral is (if $u = \cos\theta$)

$$\int_{-1}^{1} (1-u^2)\cos 2\pi nu\, du$$

$$= -\int_{-1}^{1} u^2\cos 2\pi nu\, du = -\frac{1}{\pi^2 n^2}$$

after integrating twice by parts.

If the two oscillators did not interfere, the corresponding integral is $\frac{4}{3}$, so the ratio is

$$1-\frac{3}{4(\pi n)^2}.$$

Example 5. A wire of negligible resistance lies along the z-axis and carries a varying charge $q(z, t)$ per unit length and a current $j(z, t)$. Given that the lines of force of $\boldsymbol{E}$ run radially outwards from the wire, and that those of $\boldsymbol{B}$ are circles round the wire, evaluate $E(r, z, t)$ and $B(r, z, t)$ in terms of q and j, where r denotes distance from the wire; and show that

$$\frac{\partial E}{\partial z}+\frac{\partial B}{\partial t} = 0,$$

$$\frac{\partial B}{\partial z}+\frac{1}{c^2}\frac{\partial E}{\partial t} = 0.$$

Comment on the relation between the last equation and the equation of conservation of charge on the wire.

The expression for $\boldsymbol{A}$ is

$$\boldsymbol{A}(r, z, t) = \frac{\boldsymbol{k}}{4\pi}\int_{-\infty}^{\infty}\frac{\mu_0 j(z', t-r'/c)\, dz'}{r'},$$

where $$r'^2 = r^2+(z'-z)^2.$$

The corresponding expression for the field is

$$\boldsymbol{B} = \textbf{curl}\, \boldsymbol{A} = \frac{\mu_0}{4\pi}\left(\nabla\int_{-\infty}^{\infty}\frac{j(z', t-r'/c)\, dz'}{r'}\right)\times\boldsymbol{k}$$

$$= \frac{\mu_0}{4\pi}\int\left(-\frac{1}{cr'}\frac{r\hat{\boldsymbol{r}}}{r'}\frac{\partial j}{\partial t}-\frac{r\hat{\boldsymbol{r}}j}{r'^3}\right)dz'\times\boldsymbol{k}$$

$$= -\left[\frac{\mu_0}{4\pi}\int_{-\infty}^{\infty}\left(\frac{r}{cr'^2}\frac{\partial j}{\partial t}+\frac{rj}{r'^3}\right)dz'\right]\hat{\boldsymbol{r}}\times\boldsymbol{k}$$

$$= \frac{\mu_0}{4\pi}\int_{-\infty}^{\infty}\left(\frac{r}{cr'^2}\frac{\partial j}{\partial t}+\frac{rj}{r'^3}\right)dz'\,\hat{\boldsymbol{\phi}},$$

t being understood that j and $\partial j/\partial t$ are retarded values. The scalar potential is most easily ound by noting the Lorentz condition,

$$\frac{\partial \phi}{\partial t} = -c^2 \operatorname{div} \boldsymbol{A}$$

$$= -\frac{\mu_0 c^2}{4\pi} \int_{-\infty}^{\infty} \left(\frac{(z-z')}{cr'^2} \frac{\partial j}{\partial t} + \frac{(z-z')j}{r'^3} \right) dz'$$

in the same way. Since, moreover, $j = dq/dt$, we can integrate this to give

$$\phi = -\frac{\mu_0 c^2}{4\pi} \int_{-\infty}^{\infty} \left(\frac{(z-z')}{cr'^2} j + \frac{(z-z')q}{r'^3} \right) dz'.$$

From ϕ, $\boldsymbol{A}$ the electric field can be found.
The Lorentz condition

$$\operatorname{div} \boldsymbol{A} + \frac{1}{c^2} \frac{\partial \phi}{\partial t} = 0$$

gives, since $\boldsymbol{E} = -\nabla\phi - \dfrac{\partial \boldsymbol{A}}{\partial t}$,

$$\frac{\partial \boldsymbol{B}}{\partial t} = \textbf{curl} \left(\frac{\partial \boldsymbol{A}}{\partial t} \right) = -\textbf{curl } \boldsymbol{E}.$$

If $\boldsymbol{E} = \{E_0\, 0\, 0\}$ in cylindrical polars, it readily follows that $\textbf{curl } \boldsymbol{E} = \{0\; \partial E/\partial z\; 0\}$, so that

$$\frac{\partial B}{\partial t} + \frac{\partial E}{\partial z} = 0.$$

The other Maxwell equation can be written

$$\textbf{curl } \boldsymbol{H} = \frac{1}{\mu_0} \textbf{curl } \boldsymbol{B} = \frac{\partial \boldsymbol{D}}{\partial t} = \varepsilon_0 \frac{\partial \boldsymbol{E}}{\partial t}$$

and since $\boldsymbol{B} = \{0\; B\; 0\}$ it follows that

$$\textbf{curl } \boldsymbol{B} = \left\{ -\frac{\partial B}{\partial z}\quad 0\quad \frac{1}{r} \frac{\partial}{\partial r}(rB) \right\}$$

(of which the last component must vanish). Hence

$$\frac{\partial B}{\partial z} + \frac{1}{c^2} \frac{\partial E}{\partial t} = 0.$$

The last equation can be written in the form

$$\frac{1}{\mu_0} \frac{\partial B}{\partial z} + \varepsilon_0 \frac{\partial E}{\partial t} = \frac{\partial H}{\partial z} + \frac{\partial D}{\partial t} = 0.$$

When applied to the field immediately outside the wire we can take $\boldsymbol{H}$ to be equal to the current density on the surface of the metal, and $\boldsymbol{D}$ is equal to the charge density on the surface. The equation can therefore be written in terms of the scalar values of the vectors

$$\frac{\partial J}{\partial z} + \frac{\partial \sigma}{\partial t} = 0.$$

In this form the equation denotes conservation of charge on the conductor, implying that changes in the current strength along the surface of the conductor (the z-direction) are due to corresponding variation—in time—of the charge density. It is the relevant form, for the surface, of the equation

$$\operatorname{div} \boldsymbol{J}+\partial\varrho/\partial t = 0.$$

Miscellaneous Exercises XIII

1. Verify that, in the notation of § 13 : 9,

$$R+\frac{\boldsymbol{v}\cdot\boldsymbol{S}}{c} = r\left(1-\frac{v^2}{c^2}\right)$$

and so deduce the alternative form for the field of a particle moving with uniform speed.

2. Carry out the calculation of the magnetic field mentioned in § 13.9, eqn. (13.54).

3. Show that the electric and magnetic intensities, $\boldsymbol{E}$ and $\boldsymbol{H}$, can be expressed in terms of potentials ϕ, $\boldsymbol{A}$ by the equations

$$\boldsymbol{E} = -\mathbf{grad}\ \phi-\frac{\partial \boldsymbol{A}}{\partial t}, \qquad \mu_0\boldsymbol{H} = \mathbf{curl}\ \boldsymbol{A}.$$

Derive the second-order differential equations connecting ϕ and $\boldsymbol{A}$, and show that it is possible to simplify them by taking ϕ and $\boldsymbol{A}$ to satisfy

$$\operatorname{div} \boldsymbol{A}+\frac{1}{c^2}\frac{\partial\phi}{\partial t} = 0.$$

The vector $\boldsymbol{Z}$ is defined by $\boldsymbol{Z} = \boldsymbol{a}\dfrac{e^{i\omega(r/c-t)}}{r}$,

where $\boldsymbol{a}$ is a constant vector, ω is a constant, and r is the distance from a given origin. Verify that

$$\text{(i)}\quad \phi = -\operatorname{div} \boldsymbol{Z}, \qquad \boldsymbol{A} = \frac{1}{c^2}\frac{\partial \boldsymbol{Z}}{\partial t},$$

$$\text{(ii)}\quad \phi = 0, \qquad \boldsymbol{A} = \mathbf{curl}\ \boldsymbol{Z},$$

give two solutions of Maxwell's equations *in vacuo* ($r \neq 0$). What is the relation between these solutions?

4. Distributions of charge and current of densities $\varrho(\boldsymbol{x}, t)$ and $\boldsymbol{j}(\boldsymbol{x}, t)$ in a region in which there is no dielectric or magnetic material give rise to electric and magnetic intensities $\boldsymbol{E}$ and $\boldsymbol{H}$. Prove that it is possible to express $\boldsymbol{E}$ and $\boldsymbol{H}$ in the form

$$\mu_0\boldsymbol{H} = \mathbf{curl}\ \boldsymbol{A}, \qquad \boldsymbol{E} = -\dot{\boldsymbol{A}}-\mathbf{grad}\ \phi,$$

where $\boldsymbol{A}$, ϕ satisfy the equations

$$\operatorname{div} \boldsymbol{A}+\frac{1}{c^2}\dot{\phi} = 0, \qquad \nabla^2\boldsymbol{A}-\frac{1}{c^2}\ddot{\boldsymbol{A}} = -\boldsymbol{j}(\boldsymbol{x}, t),$$

$$\nabla^2\phi-\frac{1}{c^2}\ddot{\phi} = -\varrho(\boldsymbol{x}, t)/\varepsilon_0.$$

Prove that, in a region where there is no current,

$$A = k\{f(r-ct)-rf'(r-ct)\}\frac{\cos\theta}{r^2}$$

is a solution of the wave equation determining A. In this expression k is a constant vector, $r = |r|$, $\cos\theta = z/r$, and f is any thrice-differentiable function.

5. Write down Maxwell's equations for free space and show that there is a possible solution given by

$$E = \textbf{curl curl } (k\phi),$$

$$\mu_0 H = \textbf{curl } (\dot{\phi}k/c^2),$$

where k is a fixed unit vector and ϕ is a scalar function that satisfies the wave equation $\nabla^2\phi = \ddot{\phi}/c^2$.

An electric dipole at the origin has a moment $kM(t)$ which varies with the time. Find the function ϕ for the field produced by this dipole and show that the equation of a line of electric force in a plane containing the origin and the polar axis is, in terms of polar coordinates,

$$c^{-1}\dot{M}(t-r/c)+r^{-1}M(t-r/c) = A\,\text{cosec}^2\,\theta,$$

where A is a constant, θ being measured from the direction of k.

6. Show that there is a solution of Maxwell's equations for electromagnetic waves *in vacuo* in which the components of the magnetic intensity are

$$H_x = \partial^2 S/\partial y\,\partial t, \quad H_y = -\partial^2 S/\partial x\,\partial t, \quad H_z = 0,$$

where $rS = f(ct-r)$, r is the distance from the origin, c the speed of light and f an arbitrary function.

Obtain the corresponding formulae for the components of the electric intensity; and prove that the lines of electric force are the meridian curves of the surfaces

$$\varrho\partial S/\partial\varrho = \text{constant},$$

where

$$\varrho = (x^2+y^2)^{1/2}.$$

7. Write down Maxwell's equations for the vacuum. Show that these equations are satisfied if E, H are given by

$$E = -\frac{\partial A}{\partial t} - \textbf{grad}\,\phi, \quad \mu_0 H = \textbf{curl}\,A,$$

where the scalar ϕ and the vector A are related by

$$\text{div}\,A + \frac{1}{c^2}\frac{\partial\phi}{\partial t} = 0,$$

and where ϕ and the cartesian components of A satisfy the homogeneous wave equation.

Prove that the expression

$$\frac{\partial^{m+n+s}}{\partial x^m\,\partial y^n\,\partial z^s}\left\{\frac{f(ct-r)}{r}+\frac{g(ct+r)}{r}\right\}$$

satisfies the homogeneous wave equation, where f, g are arbitrary differentiable functions, m, n, s are positive integers, and r is the distance from the origin.

It is given that

$$\phi = -\frac{\partial}{\partial z}\left\{\frac{f(ct-r)}{r}\right\}.$$

Verify that

$$A = \frac{k}{c^2}\frac{\partial}{\partial t}\left\{\frac{f(ct-r)}{r}\right\}$$

satisfies the equations, and show that this solution gives the field of a dipole of moment $kf(ct)$ situated at the origin, k being a unit vector in the direction of the z-axis.

8. Show that for an electric dipole of moment $\varepsilon_0 kf(t)$ situated at the origin, the vector potential at any point is given by

$$A = k\frac{1}{cr}\frac{\partial}{\partial t}f\left(t-\frac{r}{c}\right),$$

where k is a unit vector in the direction of the z-axis and r is the radius vector.

Determine the corresponding scalar potential, and hence find the fields E and H expressed in spherical polar coordinates.

In the case of an oscillating dipole with $f(t) = M \sin \omega t$, prove that the mean rate of energy radiation is $4\pi M^2\omega^4/(C^3\varepsilon_0 c^3)$.

9. A current j flows in a homogeneous region V of dielectric constant K and permeability μ. Assuming that the vector potential A and the scalar potential ϕ may be chosen to satisfy

$$\nabla\cdot A + \frac{\mu K}{c^2}\frac{\partial\phi}{\partial t} = 0,$$

find the equations satisfied by A and ϕ.

Assuming that a solution for A at the point (x', y', z') when $j = J(x, y, z)e^{-i\omega t}$ is of the form

$$A(x', y', z') = \mu\int_V J\frac{e^{-i(\omega t - kr)}}{r}\,dx\,dy\,dz,$$

where

$$r^2 = (x-x')^2+(y-y')^2+(z-z')^2, \qquad k = \omega(\mu K\varepsilon_0)^{1/2},$$

show that a solution for the electric field vector is

$$E(x', y', z') = \frac{ie^{-i\omega t}}{\omega K\varepsilon_0}\int_V [(J\cdot\nabla)\nabla + k^2 J]\frac{e^{ikr}}{r}\,dx\,dy\,dz,$$

where

$$\nabla = i\frac{\partial}{\partial x'} + j\frac{\partial}{\partial y'} + k\frac{\partial}{\partial z'}.$$

10. Define the electromagnetic vector potential A and the scalar potential ϕ. Determine what condition on A and ϕ will ensure that in the absence of material,

$$\frac{1}{c^2}\frac{\partial^2 A}{\partial t^2} - \nabla^2 A = j.$$

A point charge q is oscillating in free space, so that at time t its position is

$$R(t) = R_0 \cos \omega t,$$

where $R_0\omega \ll c$. Show that, for $r \gg c/\omega$,

$$B(r, t) = -\frac{q}{4\pi r^2} r\times \ddot{R}(t-r/c).$$

CHAPTER 14

THE MOTION OF CHARGED PARTICLES

14.1. Introduction

The motion of electrically charged particles in electromagnetic and gravitational fields is of interest in various studies of astronomical and engineering problems. During the past five decades research into the motion of charged particles in the terrestrial magnetic and gravitational fields has greatly improved our understanding of the structure and extent of the terrestrial magnetic field. In electronics, devices such as the magnetron were designed after theoretical studies similar to those illustrated in the examples of this chapter; in thermonuclear research, studies of magnetic bottles, wells and traps are of prime importance.

However, much care must be taken when attempts are made to generalize the results for a single particle to the case of an ionized gas, which consists of a multitude of charged particles, in a magnetic field, since the interactions between charged particles can modify or completely change the calculated (and observed) results concerning a single particle. In fact the essential feature of the motion of a charged particle in a magnetic field is the tendency of the particle to spiral around the magnetic field lines. This is regarded as the counterpart of the magnetohydrodynamic (or ionized-gas) phenomenon in which a highly (electrically) conducting material can flow freely along the lines of magnetic force but motion of the material perpendicular to the magnetic field carries the magnetic field lines with the material. [This result is similar to the classical hydrodynamic result that vorticity lines are "frozen" into a perfect fluid.]

In this chapter, we first give an account of the non-relativistic motion of a charged particle in uniform crossed electric and magnetic fields. This is followed by a brief account of the flow of charged particles when space charge density is taken into account. Finally relativistic corrections are considered. However, in this chapter we give only an elementary account of the motion in crossed electric and magnetic fields. For more detailed investigations in cases where the magnetic field is non-uniform or gravitational fields are present the reader should consult more advanced works devoted entirely to this subject.

14.2 Non-relativistic motion of an electric charge in an electromagnetic field

The fundamental equation for the motion of a particle of mass m carrying an electric charge e in crossed electric and magnetic fields $\boldsymbol{E}$ and $\boldsymbol{B}$ respectively and moving with velocity $\boldsymbol{v}$, is (in the absence of other forces)

$$m\frac{\mathrm{d}\boldsymbol{v}}{\mathrm{d}t} = e(\boldsymbol{E}+\boldsymbol{v}\times\boldsymbol{B}). \tag{14.1}$$

[The term $e\boldsymbol{v}\times\boldsymbol{B}$ on the right-hand side is, of course, the Lorentz force on the particle (see Vol. 2, p. 359).]

Clearly, since the Lorentz force is perpendicular to the velocity, the magnetic field does no work on the charge and so the changes in kinetic energy of the particle arise solely from the effect of the electric field. In fact, taking the scalar product of eqn. (14.1) with $\boldsymbol{v}$ we find

$$\frac{\mathrm{d}}{\mathrm{d}t}\left(\frac{1}{2}mv^2\right) = e\boldsymbol{E}\cdot\boldsymbol{v}. \tag{14.2}$$

In particular, when $\boldsymbol{E} = \boldsymbol{0}$ the kinetic energy, and therefore the speed v of the particle remain constant. Further, if the magnetic field $\boldsymbol{B}$ is constant, the path of the particle is a circular helix described at constant speed. This result is proved as follows:

Choose the z-axis of coordinates to be parallel to $\boldsymbol{B}$ so that

$$\boldsymbol{B} = \{0\ 0\ B\}, \quad \boldsymbol{r} = \{x\ y\ z\}.$$

The resolutes of the equation of motion are

$$m\ddot{x} = eB\dot{y}, \quad m\ddot{y} = -eB\dot{x}, \quad m\ddot{z} = 0,$$

which can be written

$$\ddot{x} = n\dot{y}, \quad \ddot{y} = -n\dot{x}, \quad \ddot{z} = 0,$$

where $n = eB/m$. Writing $\zeta = x+\mathrm{i}y$ these equations are equivalent to

$$\ddot{\zeta} = -\mathrm{i}n(\dot{x}+\mathrm{i}\dot{y}) = -\mathrm{i}n\dot{\zeta}, \quad \ddot{z} = 0$$

which integrate to give

$$\dot{\zeta} = -in(\zeta-\zeta_0), \quad \dot{z} = w,$$

where ζ_0, w are constants of integration. Integrating again

$$\zeta-\zeta_0 = a\,e^{-int}, \quad z = wt+z_0,$$

where a (complex), z_0 are constants. Writing $a = \alpha\,e^{i\delta}$, where α, δ are real, we find

$$\zeta-\zeta_0 = \alpha\,e^{-i(nt-\delta)} \tag{14.3}$$

so that

$$x-x_0 = \alpha\cos(nt-\delta), \quad y-y_0 = -\alpha\sin(nt-\delta), \quad z-z_0 = wt.$$

These are the parametric equations of a circular helix. In fact the particle moves so that the resolute $v_{\parallel}$ of its velocity parallel to B remains constant and if $e > 0$ it gyrates about the lines of magnetic force in the (clockwise) sense shown in Fig. 14.1. This gyration about the lines of magnetic force

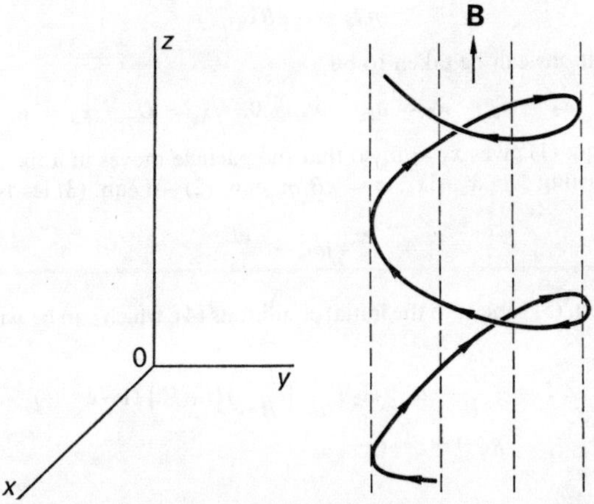

FIG. 14.1

is the fundamental characteristic of the motion of a charged particle in a magnetic field. Note also that since the speed of the particle is constant the resolute of velocity perpendicular to the magnetic field, $v_{\perp}$, is also constant and so the trajectory cuts the lines of force at a constant angle. The radius of the circular cylinder on which the spiral lies is called the *Larmor radius* of the particle; this radius is, in fact, α where α is defined in eqn. (14.3). But, from (14.3), $|\dot{\zeta}| = n\alpha$, and since $v_{\perp} = \sqrt{(\dot{x}^2+\dot{y}^2)} = |\dot{\zeta}|$, we find that the Larmor radius is

$$v_{\perp}/n = mv_{\perp}/(eB). \tag{14.4}$$

We can best describe the motion as a uniform drift along a line of magnetic force combined with a steady circular motion around that line of force. The angular velocity of rotation about the line of force is $n = eB/m$ and is called the *cyclotron frequency* (or Larmor frequency) of the charge.

In the following examples we discuss a number of phenomena which bear on problems of physical interest.

Example 1. A particle of mass m and charge e moves in a uniform magnetic field $\boldsymbol{B} = \{B\,0\,0\}$ and a uniform electric field $\boldsymbol{E} = \{0\,E\,0\}$, referred to rectangular cartesian axes $Ox_1x_2x_3$. The particle initially moves in a plane perpendicular to Ox_1. Prove that in general its path is a uniform circular motion relative to a centre which drifts with the velocity $\{0\,0\,-E/B\}$. Show also that the frequency of the circular motion is $eB/(2\pi m)$.

The equation of motion

$$m\frac{d\boldsymbol{v}}{dt} = e\boldsymbol{E}+e\boldsymbol{v}\times\boldsymbol{B}$$

has resolutes

$$m\ddot{x}_1 = 0, \tag{1}$$

$$m\ddot{x}_2 = eE+eB\dot{x}_3, \tag{2}$$

$$m\ddot{x}_3 = -eB\dot{x}_2. \tag{3}$$

The initial conditions can be taken to be

$$x_1 = a_1,\quad x_2 = a_2,\quad x_3 = a_3,\quad \dot{x}_1 = 0,\quad \dot{x}_2 = U,\quad \dot{x}_3 = V,\ \text{at}\ t = 0. \tag{4}$$

Integration of eqn. (1) gives $x_1 = a_1$ so that the particle moves in a fixed plane perpendicular to Ox_1. Writing $\zeta = x_2+\mathrm{i}x_3$, $\omega = eB/m$, eqn. (2) $+\mathrm{i}$ eqn. (3) leads to

$$\ddot{\zeta}+\mathrm{i}\omega\dot{\zeta} = \frac{eE}{m}. \tag{5}$$

The solution of eq. (5) subject to the initial conditions (4), which can be written $\zeta = a_2+\mathrm{i}a_3$, $\dot{\zeta} = U+\mathrm{i}V$, is

$$\zeta = -\frac{\mathrm{i}Et}{B}+a_1+\mathrm{i}a_2+\frac{1}{\omega}\left(\frac{E}{B}-\mathrm{i}U+V\right)(1-e^{-\mathrm{i}\omega t}),$$

i.e.

$$\zeta-\zeta_P = Re^{-\mathrm{i}(\omega t-\pi+\phi)}, \tag{6}$$

where

$$\zeta_P = \left(a_1+\frac{E}{\omega B}+\frac{V}{\omega}\right)+\mathrm{i}\left(a_2-\frac{U}{\omega}-\frac{Et}{B}\right),\qquad Re^{-\mathrm{i}\phi} = \frac{1}{\omega}\left(\frac{E}{B}+V-\mathrm{i}U\right),$$

so that R, ϕ are real constants.

Equation (6) implies that the point ζ is a fixed distance R from the point ζ_P which itself drifts with velocity $-E/B$ parallel to Ox_3. Further the vector ζ_P rotates clockwise with angular velocity ω. Therefore the motion is as described, the frequency of the angular motion being $\omega/(2\pi) = eB/(2\pi m)$.

Example 2. A beam of electrons passes through a small aperture into a uniform magnetic field of strength B. If the electrons have a speed v at the aperture and are moving in directions nearly parallel to the magnetic field, prove that the beam will focus at a distance $2\pi mv/(eB)$ from the aperture, where $-e$ and m denote the charge and mass of the electron. (Neglect interactions between the electrons.)

Again we use rectangular cartesian coordinates $Ox_1x_2x_3$ but this time take $\boldsymbol{B} = \{0\ 0\ B\}$. Then the equation of motion for a single electron is

$$m\frac{\mathrm{d}\boldsymbol{v}}{\mathrm{d}t} = -e\boldsymbol{v}\times\boldsymbol{B}$$

with resolutes

$$\ddot{x}_1+\omega\dot{x}_2 = 0, \quad \ddot{x}_2-\omega\dot{x}_1 = 0, \quad \ddot{x}_3 = 0,$$

where $\omega = eB/m$. Integrating these equations as in Example 1 above we find that the coordinates of the particle which starts from $\{0, 0, 0\}$ with speed $\{U\ V\ v\}$, where U, V are small compared with v, are

$$x_1+\mathrm{i}x_2 = +\frac{1}{\omega}(\mathrm{i}U-V)(1-\mathrm{e}^{\mathrm{i}\omega t}), \tag{1}$$

$$x_3 = vt. \tag{2}$$

Then

$$x_1-\mathrm{i}x_2 = -\frac{1}{\omega}(\mathrm{i}U+V)(1-\mathrm{e}^{\mathrm{i}\omega t}). \tag{3}$$

Multiplying corresponding sides of eqns. (1) and (3) we find

$$x_1^2+x_2^2 = 2(U^2+V^2)(1-\cos\omega t)/\omega^2.$$

The beam will focus where $x_1^2+x_2^2$ is least, i.e. where $\cos\omega t = 1$. This focussing first takes place when $t = 2\pi/\omega$ at which instant the particles are distant $2\pi v/\omega = 2\pi vm/(eB)$ from the aperture.

Example 3. An electron of negligible mass carries a charge e and moves with uniform velocity $\boldsymbol{v}$ in a medium which offers a resistance $k\boldsymbol{v}$ to its motion. The motion is driven by uniform electric and magnetic fields

$$\boldsymbol{E} = \{E\ \ 0\ \ 0\}, \quad \boldsymbol{B} = \{0\ \ B\ \ 0\}.$$

Calculate $\boldsymbol{v}$ in terms of e, k, $\boldsymbol{E}$, and $\boldsymbol{B}$.

A conducting medium may be regarded as composed of a large number n of electrons per unit volume, whose motion is resisted (according to the above law, with k given) by a *fixed* framework of positively charged particles, the total charge-density of the medium being zero. On the assumption that the magnetic field due to the current j is negligible, calculate the conductivities σ_1 and σ_2 in the relation

$$j = \{\sigma_1 E\ \ 0\ \ \sigma_2 E\}.$$

When the conductor is bounded by planes z = constant, initial currents in the z-direction establish surface charges on the bounding planes which give rise to a z-component of electric field. Calculate the strength of this field in the final steady state when the z-component of current has been reduced to zero. Show also that the corresponding conductivity σ_3 in the relation

$$j = \{\sigma_3 E\ \ 0\ \ 0\}$$

satisfies the equation

$$\sigma_3 = \sigma_1+\frac{\sigma_2^2}{\sigma_1}.$$

The equation of motion of the electron is

$$m\frac{\mathrm{d}\boldsymbol{v}}{\mathrm{d}t} = e\boldsymbol{E}-k\boldsymbol{v}+e\boldsymbol{v}\times\boldsymbol{B}. \tag{1}$$

When the electron moves with constant velocity, $d\boldsymbol{v}/dt = \boldsymbol{0}$, and eqn. (1) reduces to

$$e\boldsymbol{v}\times\boldsymbol{B}-k\boldsymbol{v} = -e\boldsymbol{E}. \tag{2}$$

Taking scalar product of (2) with $\boldsymbol{B}$ gives

$$-k\boldsymbol{v}\cdot\boldsymbol{B} = -e\boldsymbol{E}\cdot\boldsymbol{B}. \tag{3}$$

Also taking vector product of (2) with $\boldsymbol{B}$ gives

$$e(\boldsymbol{v}\times\boldsymbol{B})\times\boldsymbol{B}-k\boldsymbol{v}\times\boldsymbol{B} = -e\boldsymbol{E}\times\boldsymbol{B},$$

i.e.

$$e\{(\boldsymbol{v}\cdot\boldsymbol{B})\boldsymbol{B}-B^2\boldsymbol{v}\}-k\boldsymbol{v}\times\boldsymbol{B} = -e\boldsymbol{E}\times\boldsymbol{B}.$$

Then using eqn. (2) to substitute for $\boldsymbol{v}\times\boldsymbol{B}$ and eqn. (3) to substitute for $\boldsymbol{v}\cdot\boldsymbol{B}$, we have finally

$$\boldsymbol{v} = \frac{e}{k^2+e^2B^2}\left\{e\boldsymbol{E}\times\boldsymbol{B}+k\boldsymbol{E}+\frac{e^2}{k}(\boldsymbol{E}\cdot\boldsymbol{B})\boldsymbol{B}\right\}. \tag{4}$$

The current density vector $\boldsymbol{j}$ is given by

$$\boldsymbol{j} = ne\boldsymbol{v}, \tag{5}$$

see Vol. 1, p. 153. With the given expression for $\boldsymbol{E}$, $\boldsymbol{B}$ we find

$$\boldsymbol{j} = \frac{ne^2E}{k^2+e^2B^2}\{k \quad 0 \quad eB\}.$$

Hence

$$\sigma_1 = \frac{kne^2}{k^2+e^2B^2}, \qquad \sigma_2 = \frac{ne^3B}{k^2+e^2B^2}.$$

When the z-component of current has been reduced to zero, let the z-component of electric field be E_3. Then eqns. (4) and (5) now give

$$\boldsymbol{j} = \frac{ne^2}{k^2+e^2B^2}\{kE-eBE_3 \quad 0 \quad eBE+kE_3\};$$

(he vanishing of the z-component implies that

$$E_3 = -eBE/k.$$

Therefore

$$\boldsymbol{j} = \{j_x \quad 0 \quad 0\},$$

where

$$\begin{aligned} j_x &= \frac{ne^2(kE+e^2B^2E/k)}{k^2+e^2B^2} \\ &= \left\{\frac{kne^2}{k^2+e^2B^2}+\left(\frac{ne^3B}{k^2+e^2B^2}\right)^2\Big/\left(\frac{kne^2}{k^2+e^2B^2}\right)\right\}E \\ &= \sigma_3 E. \end{aligned}$$

Therefore

$$\sigma_3 = \sigma_1+\frac{\sigma_2^2}{\sigma_1}.$$

Example 4. A particle of mass m and charge q, initially at rest at the origin, is acted on by a constant electric field $\boldsymbol{i}E$ and a constant magnetic field $\boldsymbol{k}B$, where $\boldsymbol{i}$ and $\boldsymbol{k}$ are unit vectors along the x-and z-axes. The particle is subjected to a resistance $m\lambda\boldsymbol{v}$, where $\boldsymbol{v}$ is the velocity and λ is a constant. Show that the subsequent motion is given by

$$x+\mathrm{i}y = \frac{qE}{pm}t-\frac{qE}{mp^2}(1-\mathrm{e}^{-pt}), \quad p = \lambda+\mathrm{i}\frac{qB}{m}, \quad z = 0,$$

and determine the direction and magnitude of the terminal velocity.

As in Example 3 above, the equation of motion of the particle is

$$\frac{\mathrm{d}\boldsymbol{v}}{\mathrm{d}t} = \frac{q\boldsymbol{E}}{m}-\lambda\boldsymbol{v}+\frac{q\boldsymbol{v}}{m}\times\boldsymbol{B}$$

with cartesian resolutes

$$\ddot{x}+\lambda\dot{x}-\frac{qB}{m}\dot{y} = \frac{qE}{m}, \tag{1}$$

$$\ddot{y}+\lambda\dot{y}+\frac{qB}{m}\dot{x} = 0, \tag{2}$$

$$\ddot{z}+\lambda\dot{z} = 0. \tag{3}$$

The initial conditions are $x = y = z = 0$, $\dot{x} = \dot{y} = \dot{z} = 0$, at $t = 0$.

Equation (1)+i eqn. (2) gives

$$\ddot{\zeta}+\left(\lambda+\frac{\mathrm{i}qB}{m}\right)\dot{\zeta} = \frac{qE}{m},$$

where $\zeta = x+\mathrm{i}y$. The solution of this differential equation (subject to the initial conditions $\dot{\zeta} = 0 = \zeta$ at $t = 0$) is

$$\zeta = \frac{qEt}{pm}-\frac{qE}{mp^2}(1-\mathrm{e}^{-pt}),$$

where $p = \lambda+\mathrm{i}qB/m$, as required. Equation (3) integrates to give $z = 0$.

As $t \to \infty$,

$$\dot{x}+\mathrm{i}\dot{y} \to \frac{qE}{pm} = \frac{qE/(\lambda-\mathrm{i}qB/m)}{m(\lambda^2+q^2B^2/m^2)}.$$

Hence the terminal velocity has resolutes

$$\left\{\frac{\lambda qE}{K} \quad -\frac{q^2EB/m}{K} \quad 0\right\},$$

where $K = m(\lambda^2+q^2B^2/m^2)$.

Example 5. A certain rectifying device consists of two coaxial circular cylindrical conductors of radii a and $b (a < b)$. A potential difference is maintained between the conductors, and a magnetic field parallel to the axis is applied; the space between the conductors is evacuated. Electrons, of mass m (assumed constant) and charge q, are released from the inner cylinder with negligible velocity. Assuming that the electric potential ϕ and the magnetic flux density B are functions of the axial distance r, prove that

$$\tfrac{1}{2}m(\dot{r}^2+r^2\dot{\theta}^2) = q\{\phi(a)-\phi(r)\} \quad \text{and} \quad mr^2\dot{\theta} = -q\int_a^r sB(s)\,\mathrm{d}s,$$

where θ is the longitude.

Deduce that, if the electrons reach the outer cylinder, the potential difference between the cylinders must be at least as great as

$$|q|\, N^2/(8\pi^2 mb^2),$$

where N is the magnetic flux between the cylinders.
[Neglect interactions between the electrons.]

We use cylindrical polar coordinates (r, θ, z) with the origin on the common axis of the conductors and Oz along the axis. Then the longitudinal (transverse) and axial resolutes of the equation of motion are

$$\frac{m}{r}\frac{\mathrm{d}}{\mathrm{d}t}(r^2\dot{\theta}) = -qB(r)\frac{\mathrm{d}r}{\mathrm{d}t}, \tag{1}$$

$$m\ddot{z} = 0. \tag{2}$$

Equation (1), written in the form

$$m\frac{\mathrm{d}}{\mathrm{d}t}(r^2\dot{\theta}) = -qrB(r)\frac{\mathrm{d}r}{\mathrm{d}t},$$

integrates (w.r. to t) to give

$$mr^2\dot{\theta} = -q\int_a^r sB(s)\,\mathrm{d}s \tag{3}$$

on using the conditions $\dot{r} = 0 = \dot{\theta}$ when $r = a$ (following from negligible velocity of emission).

Equation (2), combined with the initial condition $z = 0 = \dot{z}$ at emission, implies that $z =$ constant and so each electron moves in a plane perpendicular to the axis. Then the energy equation (14.2) implies that the gain of kinetic energy $\frac{1}{2}mv^2$ is equal to the loss of potential energy $q\{\phi(a)-\phi(r)\}$. Expressed in cylindrical polar coordinates this gives

$$\tfrac{1}{2}m(\dot{r}^2+r^2\dot{\theta}^2) = q\{\phi(a)-\phi(r)\}. \tag{4}$$

From eqns. (3) and (4) we find

$$\dot{r}^2 = \frac{2q}{m}\{\phi(a)-\phi(r)\}-\frac{q^2}{m^2r^2}\left\{\int_a^r sB(s)\,\mathrm{d}s\right\}^2. \tag{5}$$

For the electrons to reach the outer cylinder $\dot{r}^2 > 0$ for $a < r < b$ and in particular $\dot{r}^2 \geqslant 0$ when $r = b$, i.e.

$$\frac{2q}{m}\{\phi(a)-\phi(b)\} \geqslant \frac{q^2}{m^2b^2}\left\{\int_a^b sB(s)\,\mathrm{d}s\right\}^2. \tag{6}$$

But the magnetic flux, N, between the cylinders is given by

$$N = 2\pi\int_a^b s\,B(s)\,\mathrm{d}s$$

and so the inequality (6) becomes

$$-q\{\phi(b)-\phi(a)\} \geqslant q^2N^2/(8\pi^2mb^2)$$

which, since q is negative, reduces to

$$\phi(b)-\phi(a) \geqslant |q|\, N^2/(8\pi^2 mb^2)$$

which is the required relation.

Example 6. The equation of motion of a charged particle moving in uniform constant electric and magnetic fields $\boldsymbol{E}$ and $\boldsymbol{B}$ is

$$\ddot{\boldsymbol{r}} = \boldsymbol{E}+\dot{\boldsymbol{r}}\times\boldsymbol{B} \tag{1}$$

where $\boldsymbol{r}$ is the position vector and dots denote differentiation with respect to time.

The particle is projected from the origin with velocity $\boldsymbol{v}$; $\boldsymbol{E}$ is parallel to $\boldsymbol{v}$, but perpendicular to $\boldsymbol{B}$. Show that after a time t,

$$\dot{\boldsymbol{r}} = \boldsymbol{E}t+\boldsymbol{r}\times\boldsymbol{B}+\boldsymbol{v}, \tag{2}$$

$$\dot{\boldsymbol{r}}\cdot\dot{\boldsymbol{r}} = \boldsymbol{v}\cdot\boldsymbol{v}+2\boldsymbol{E}\cdot\boldsymbol{r}, \tag{3}$$

$$\boldsymbol{r}\cdot\boldsymbol{B} = 0, \tag{4}$$

$$\ddot{\boldsymbol{r}}+B^2\boldsymbol{r} = \boldsymbol{E}+\boldsymbol{v}\times\boldsymbol{B}+\boldsymbol{E}\times\boldsymbol{B}t. \tag{5}$$

Hence deduce that

$$\boldsymbol{r} = (\boldsymbol{E}+\boldsymbol{v}\times\boldsymbol{B})\,(1-\cos Bt)/B^2+(B^2\boldsymbol{v}+\boldsymbol{B}\times\boldsymbol{E})\,(\sin Bt)/B^3+(\boldsymbol{E}\times\boldsymbol{B})t/B^2\,.$$

Equation (1) integrates at once to give

$$\dot{\boldsymbol{r}} = \boldsymbol{E}t+\boldsymbol{r}\times\boldsymbol{B}+\boldsymbol{C},$$

where $\boldsymbol{C}$ is a constant vector. The initial conditions $\boldsymbol{r} = \boldsymbol{0}$, $\dot{\boldsymbol{r}} = \boldsymbol{v}$ give $\boldsymbol{C} = \boldsymbol{v}$ and so eqn. (2) follows.

Taking the scalar product of both sides of eqn. (1) with $\dot{\boldsymbol{r}}$ we have

$$\ddot{\boldsymbol{r}}\cdot\dot{\boldsymbol{r}} = \boldsymbol{E}\dot{\boldsymbol{r}},$$

i.e.

$$\frac{\mathrm{d}}{\mathrm{d}t}\left(\frac{1}{2}\,\dot{\boldsymbol{r}}\cdot\dot{\boldsymbol{r}}\right) = \frac{\mathrm{d}}{\mathrm{d}t}\,(\boldsymbol{E}\cdot\boldsymbol{r}).$$

Integrating we find

$$\left[\tfrac{1}{2}\dot{\boldsymbol{r}}\cdot\dot{\boldsymbol{r}}\right]_0^t = \left[\boldsymbol{E}\cdot\boldsymbol{r}\right]_0^t,$$

i.e.,

$$\dot{\boldsymbol{r}}\cdot\dot{\boldsymbol{r}} = \boldsymbol{v}\cdot\boldsymbol{v}+2\boldsymbol{E}\cdot\boldsymbol{r}. \tag{3}$$

Taking the scalar product of both sides of eqn. (2) with $\boldsymbol{B}$, remembering that $\boldsymbol{r}\times\boldsymbol{B}$ is perpendicular to $\boldsymbol{B}$ by definition and that $\boldsymbol{E}$ and $\boldsymbol{v}$ are given to be perpendicular to $\boldsymbol{B}$, we find

$$\dot{\boldsymbol{r}}\cdot\boldsymbol{B} = 0.$$

Therefore

$$\frac{\mathrm{d}}{\mathrm{d}t}\,(\boldsymbol{r}\cdot\boldsymbol{B}) = 0$$

or $\boldsymbol{r}\cdot\boldsymbol{B} = A$, where A is a constant scalar. Since $\boldsymbol{r} = \boldsymbol{0}$ at $t = 0$, it follows that $A = 0$ and eqn. (4) holds.

Now substitute $\dot{\boldsymbol{r}}$ from eqn. (2) into the r.h. side of eqn. (1). Then

$$\begin{aligned}\ddot{\boldsymbol{r}} &= \boldsymbol{E}+(\boldsymbol{E}t+\boldsymbol{r}\times\boldsymbol{B}+\boldsymbol{v})\times\boldsymbol{B} = \boldsymbol{E}+\boldsymbol{E}\times\boldsymbol{B}t+(\boldsymbol{r}\times\boldsymbol{B})\times\boldsymbol{B}+\boldsymbol{v}\times\boldsymbol{B}\\ &= \boldsymbol{E}+\boldsymbol{E}\times\boldsymbol{B}t-B^2\boldsymbol{r}+(\boldsymbol{B}\cdot\boldsymbol{r})\boldsymbol{B}+\boldsymbol{v}\times\boldsymbol{B},\end{aligned}$$

i.e.

$$\ddot{\boldsymbol{r}}+B^2\boldsymbol{r} = \boldsymbol{E}+\boldsymbol{v}\times\boldsymbol{B}+\boldsymbol{E}\times\boldsymbol{B}t \tag{5}$$

on using eqn. (4).

The solution of eqn. (5) is

$$\boldsymbol{r} = \boldsymbol{r}_c + \boldsymbol{r}_p, \tag{6}$$

where

$$\ddot{\boldsymbol{r}}_c + B^2\boldsymbol{r}_c = \boldsymbol{0},$$

i.e.

$$\boldsymbol{r}_c = \boldsymbol{L}\cos Bt + \boldsymbol{M}\sin Bt, \tag{7}$$

$\boldsymbol{L}$, $\boldsymbol{M}$, being arbitrary vectors. The particular integral of (5) is clearly

$$\boldsymbol{r}_p = (\boldsymbol{E} + \boldsymbol{v}\times\boldsymbol{B} + \boldsymbol{E}\times\boldsymbol{B}t)/B^2. \tag{8}$$

Substituting from eqns. (7), (8) in eqn. (6) and using the initial conditions $\boldsymbol{r} = \boldsymbol{0}$, $\dot{\boldsymbol{r}} = \boldsymbol{v}$ leads to the required answer.

Example 7. A particle, of mass m carrying a charge e, moves *in vacuo* in an electric field $\boldsymbol{E}$ and a magnetic field with magnetic flux density vector $\boldsymbol{B}$. Referred to a system of rectangular cartesian axes $Oxyz$, $\boldsymbol{E} = \{E_0x\, E_0y\, 0\}$, $\boldsymbol{B} = \{0\, 0\, B_0\}$, where E_0, B_0 are constants. Show that the velocity of the particle parallel to Oz remains constant. Show also that, if $e^2B_0^2 > 4meE_0$, the distance of the particle from the axis Oz remains finite.

If this is so and the particle is initially at rest at the point $\{a\, 0\, 0\}$, find the maximum distance of the particle from the origin O during the subsequent motion.

Referred to rectangular cartesian axes $Oxyz$ the resolutes of the equation of motion of the particle P are

$$m\ddot{x} = eE_0x + eB_0\dot{y}, \tag{1}$$

$$m\ddot{y} = eE_0y - eB_0\dot{x}, \tag{2}$$

$$m\ddot{z} = 0. \tag{3}$$

Equation (3) integrates at once to give $\dot{z}$ = constant, implying that the velocity of P parallel to Oz remains constant. Writing $\zeta = x + iy$ and taking eqn. (1) + i eqn. (2) gives

$$m\ddot{\zeta} + \mathrm{i}eB_0\dot{\zeta} - eE_0\zeta = 0. \tag{4}$$

The general solution of eqn. (4) is

$$\zeta = P\mathrm{e}^{\omega_1 t} + Q\mathrm{e}^{\omega_2 t}$$

where P, Q are complex arbitrary constants and ω_1, ω_2 are the roots of the auxiliary equation

$$m\omega^2 + \mathrm{i}eB_0\omega - eE_0 = 0,$$

i.e.

$$\omega_1, \omega_2 = \{-\mathrm{i}eB_0 \pm \sqrt{(4meE_0 - e^2B_0^2)}\}/(2m).$$

If ω_1, ω_2 are purely imaginary, then $|\zeta|$ is bounded and so $\sqrt{(x^2+y^2)}$ remains finite. Therefore the distance of P from Oz remains finite if

$$e^2B_0^2 > 4meE_0.$$

In this case, and when the particle is initially at rest at $\{a\, 0\, 0\}$, eqn. (3) integrates to give $z = 0$. Also

$$\zeta = P\mathrm{e}^{\mathrm{i}\varkappa_1 t} + Q\mathrm{e}^{\mathrm{i}\varkappa_2 t}$$

where

$$\varkappa_1, \varkappa_2 = \{-eB_0 \pm \sqrt{(e^2B_0^2 - 4meE_0)}\}/(2m)$$

and $\zeta = a$, $\dot{\zeta} = 0$ at $t = 0$. Therefore

$$P+Q = a, \quad i\varkappa_1 P + i\varkappa_2 Q = 0,$$

whence

$$\zeta = \frac{a}{\varkappa_2-\varkappa_1}(\varkappa_2 e^{i\varkappa_1 t} - \varkappa_1 e^{i\varkappa_2 t}).$$

In this case

$$x^2+y^2 = \zeta\zeta^* = \frac{a^2\{\varkappa_2^2+\varkappa_1^2-2\varkappa_1\varkappa_2\cos(\varkappa_1-\varkappa_2)t\}}{(\varkappa_1-\varkappa_2)^2}.$$

The greatest value of $r = \sqrt{(x^2+y^2)}$ occurs when $\cos(\varkappa_1-\varkappa_2)t = -1$ and in fact

$$r_{\max} = |a(\varkappa_1+\varkappa_2)/(\varkappa_1-\varkappa_2)|$$

$$= \left|\frac{aeB_0}{(e^2B_0^2-4meE_0)^{\frac{1}{2}}}\right|.$$

Example 8. Find the equation of motion in cylindrical polars r, ϕ, z if a particle of mass m carrying a charge e is moving in the magnetic field due to a constant current I in a long straight wire along the axis of z. Find $\dot{r}^2$ as a function of r and, by sketching this function, prove that the distance of the particle from the wire oscillates between fixed upper and lower bounds.

In the equation of motion

$$m\dot{\boldsymbol{v}} = e(\boldsymbol{E}+\boldsymbol{v}\times\boldsymbol{B}), \tag{1}$$

$$\boldsymbol{B} = \left\{0 \quad \frac{\mu_0 I}{2\pi r} \quad 0\right\}, \qquad \boldsymbol{v} = \{\dot{r} \quad r\dot{\phi} \quad \dot{z}\}, \qquad \boldsymbol{E} = \boldsymbol{0},$$

$$\dot{\boldsymbol{v}} = \left\{\ddot{r}-r\dot{\phi}^2 \quad \frac{1}{r}\frac{d}{dt}(r^2\dot{\phi}) \quad \ddot{z}\right\}.$$

Therefore the resolutes of eqn. (1) are

$$\ddot{r}-r\dot{\phi}^2 = -\frac{\mu_0 eI}{2\pi m}\frac{\dot{z}}{r}, \tag{2}$$

$$\frac{1}{r}\frac{d}{dt}(r^2\dot{\phi}) = 0, \tag{3}$$

$$\ddot{z} = \frac{\mu_0 eI}{2\pi m}\frac{\dot{r}}{r}. \tag{4}$$

The energy integral gives

$$\dot{r}^2+(r\dot{\phi})^2+\dot{z}^2 = a \quad \text{(constant)} \tag{5}$$

and the angular momentum integral [the integral of eqn. (3)] is

$$r^2\dot{\phi} = h. \tag{6}$$

The first integral of eqn. (4) is

$$\dot{z} = \frac{\mu_0 eI}{2\pi m}\ln\left(\frac{r}{b}\right), \tag{7}$$

where b is constant. Therefore eqn. (5) gives [using eqns. (6), (7)]

$$\dot{r}^2 = a - \frac{h^2}{r^2} - \frac{\mu_0^2 e^2 I^2}{4\pi^2 m^2}\left[\ln\left(\frac{r}{b}\right)\right]^2 = f(r), \quad \text{say.} \tag{8}$$

Then

$$\frac{df}{dr} = \frac{1}{r}\left[\frac{2h^2}{r^2} - \frac{\mu_0^2 e^2 I^2}{2\pi^2 m^2}\ln\left(\frac{r}{b}\right)\right].$$

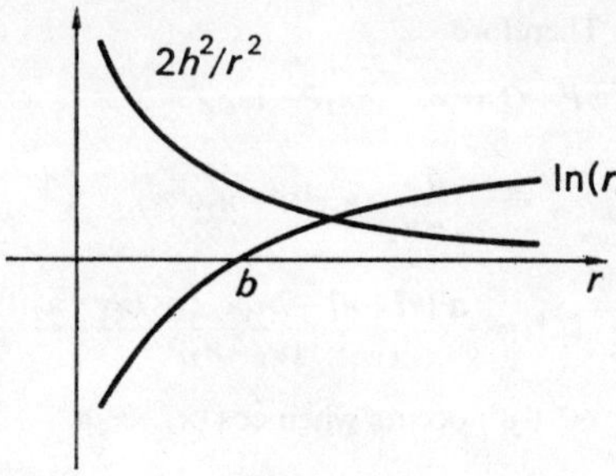

FIG. 14.2

The first term in the bracket is a strictly decreasing function of r, the second term an increasing function, see Fig. 14.2.

Therefore $\mathrm{d}f/\mathrm{d}r$ has only one zero which must be a maximum of f ($\mathrm{d}f/\mathrm{d}r$ changes from +ve to −ve there). The conditions of projection imply that $\dot{r}^2 \geqslant 0$ at, say, $r = R$.

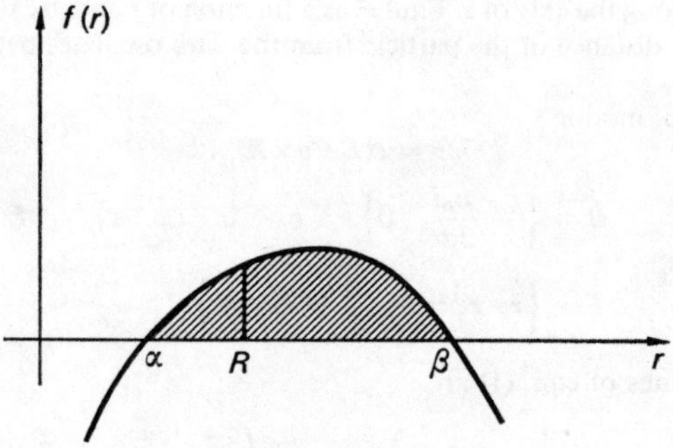

FIG. 14.3

When $\dot{r}^2 > 0$ *at* $r = R$, $f(r) > 0$ there. Also, as $r \to 0$, $f(r) > -\infty$; and as $r \to \infty$, $f(r) \to -\infty$. The graph of $f(r)$ is therefore as sketched in Fig. 14.3, and the shaded portion corresponds to the possible positions of the particle which must oscillate between α and β, the time of oscillation being finite since

$$\int_{\alpha}^{\beta} \frac{\mathrm{d}r}{\sqrt{\{f(r)\}}}$$

converges.

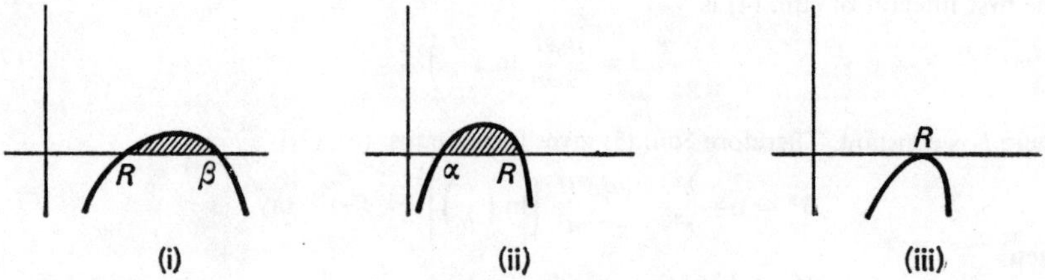

FIG. 14.4

When $\dot{r}^2 = 0$ *at* $r = R$, we may have one of the following cases shown in Fig. 14.4.

(i) the particle oscillates between R and β,
(ii) the particle oscillates between α and R,
(iii) r remains constant.

Example 9. The use of the Lagrangian

Show that the equation of motion of a charged particle may be written

$$\frac{d}{dt}\left(\frac{\partial L}{\partial \dot{x}_\alpha}\right) = \frac{\partial L}{\partial x_\alpha}, \qquad (\alpha = 1, 2, 3)$$

where

$$L = \tfrac{1}{2}mv^2 - e\phi + e\boldsymbol{v}\cdot\boldsymbol{A},$$

and where $\boldsymbol{A}$ and ϕ are the potentials of the external field, x_α are rectangular cartesian coordinates, and d/dt operating on a field variable denotes the time derivative following the motion of the particle.

If $\boldsymbol{E} = \boldsymbol{0}$ and $\boldsymbol{B}$ is uniform, show that the particle moves in a helix.

When the "generalized" coordinates x_α are in fact the cartesian coordinates of the particle, then Lagrange's equations of motion in the form

$$\frac{d}{dt}\left(\frac{\partial T}{\partial \dot{x}_\alpha}\right) - \frac{\partial T}{\partial x_\alpha} = X_\alpha \tag{1}$$

have the rectangular resolutes of the force acting on the particle as the right-hand sides. In mechanical systems under the action of conservative forces the components are the derivatives

$$X_\alpha = -\partial V/\partial x_\alpha \tag{2}$$

of the potential energy. When the forces depend on the velocities $\dot{x}_\alpha$ in a special manner the usual form of the equations of motion can be retained. This is possible when

$$X_\alpha = \frac{d}{dt}\left(\frac{\partial U}{\partial \dot{x}_\alpha}\right) - \frac{\partial U}{\partial x_\alpha}. \tag{3}$$

This reduces to (2) if $\partial U/\partial \dot{x}_\alpha = 0$.] The Lagrangian function is defined as

$$L = T - U$$

so that (1) becomes the usual form

$$\frac{d}{dt}\left(\frac{\partial L}{\partial \dot{x}_\alpha}\right) - \frac{\partial L}{\partial x_\alpha} = 0. \tag{4}$$

The Lorentz force acting on a charged particle can, in fact, be put into the for m (3) Expressed in suffix notation the Lorentz force is

$$X = e(\boldsymbol{E} + \boldsymbol{v} \times \boldsymbol{B})_\alpha = eE_\alpha + \varepsilon_{\alpha\beta\gamma}\dot{x}_\beta B_\gamma.$$

We now substitute the vector and scalar potentials

$$\boldsymbol{E} = -\mathbf{grad}\,\phi - \frac{\partial \boldsymbol{A}}{\partial t}, \qquad \boldsymbol{B} = \mathbf{curl}\,\boldsymbol{A},$$

i.e.,

$$E_\alpha = -\frac{\partial \phi}{\partial x_\alpha} - \frac{\partial A_\alpha}{\partial t}, \qquad B_\alpha = \varepsilon_{\alpha\delta\varepsilon}\frac{\partial A_\varepsilon}{\partial x_\delta}.$$

Then

$$X_\alpha = e\left\{-\frac{\partial\phi}{\partial x_\alpha}-\frac{\partial A_\alpha}{\partial t}+\varepsilon_{\alpha\beta\gamma}\varepsilon_{\gamma\delta\varepsilon}\dot{x}_\beta\frac{\partial A_\varepsilon}{\partial x_\delta}\right\}$$

$$= e\left\{-\frac{\partial\phi}{\partial x_\alpha}-\frac{\partial A_\alpha}{\partial t}+(\delta_{\alpha\delta}\delta_{\beta\varepsilon}-\delta_{\alpha\varepsilon}\delta_{\delta\beta})\dot{x}_\beta\frac{\partial A_\varepsilon}{\partial x_\delta}\right\}$$

$$= e\left\{-\frac{\partial\phi}{\partial x_\alpha}-\frac{\partial A_\alpha}{\partial t}+\dot{x}_\beta\frac{\partial A_\beta}{\partial x_\alpha}-\dot{x}_\beta\frac{\partial A_\alpha}{\partial x_\beta}\right\}.$$

Since A_α depends upon x_α and t we can write

$$\frac{dA_\alpha}{dt} = \frac{\partial A_\alpha}{\partial t}+\dot{x}_\beta\frac{\partial A_\alpha}{\partial x_\beta}.$$

Also

$$\frac{\partial}{\partial x_\alpha}(\dot{x}_\beta A_\beta) = \dot{x}_\beta\frac{\partial A_\beta}{\partial x_\alpha}.$$

Therefore

$$X_\alpha = e\left\{-\frac{\partial}{\partial x_\alpha}(\phi-\dot{x}_\beta A_\beta)-\frac{dA_\alpha}{dt}\right\}.$$

If we now introduce $U = e(\phi-\dot{x}_\beta A_\beta)$, we find

$$\frac{d}{dt}\left(\frac{\partial U}{\partial \dot{x}_\alpha}\right) = -\frac{dA_\alpha}{dt}.$$

Therefore

$$X_\alpha = e\left\{-\frac{\partial U}{\partial x_\alpha}+\frac{d}{dt}\left(\frac{\partial U}{\partial \dot{x}_\alpha}\right)\right\}.$$

The equation of motion of the particle can thus be put into the Lagrangian form (4) with

$$L = T-U = \tfrac{1}{2}mv^2-e\phi+e\dot{x}_\beta A_\beta = \tfrac{1}{2}mv^2-e\phi+e(\boldsymbol{v}\cdot\boldsymbol{A}).$$

For the uniform fields $\boldsymbol{E} = \boldsymbol{0}$, $\boldsymbol{B} = B\boldsymbol{k}$ we take $\phi = 0$, $\boldsymbol{A} = \tfrac{1}{2}B(\boldsymbol{k}\times\boldsymbol{r})$.

Then $$L = \tfrac{1}{2}mv^2+\tfrac{1}{2}eB(\boldsymbol{v}\cdot\boldsymbol{k}\times\boldsymbol{r}) = \tfrac{1}{2}m(\dot{x}_1^2+\dot{x}_2^2+\dot{x}_3^2)^2+\tfrac{1}{2}eB(x_1\dot{x}_2-x_2\dot{x}_1)$$

and the equations of motion are

$$\frac{d}{dt}(m\dot{x}_1-\tfrac{1}{2}eBx_2)-\tfrac{1}{2}eB\dot{x}_2 = 0,$$

$$\frac{d}{dt}(m\dot{x}_2+\tfrac{1}{2}eBx_1)+\tfrac{1}{2}eB\dot{x}_1 = 0,$$

$$\frac{d}{dt}(m\dot{x}_3) = 0,$$

i.e. $$\ddot{x}_1-(eB/m)\dot{x}_2 = 0, \quad \ddot{x}_2+(eB/m)\dot{x}_1 = 0, \quad \ddot{x}_3 = 0.$$

These equations are the same as those given on p. 570 and lead to eqn. (14.3) and correspond to motion on a helix.

Exercises 14.2

1. A charged particle of mass m and charge e moves under the influence of a uniform electric field of intensity E parallel to the x-axis and a uniform magnetic field of intensity B parallel to the z-axis. The particle starts from rest at the origin at time $t = 0$; show that at time t its coordinates are given by

$$x = \frac{E}{\omega B}(1-\cos \omega t), \quad y = -\frac{E}{\omega B}(\omega t - \sin \omega t), \quad z = 0,$$

where $\omega = eB/m$, and show that the speed of the particle is given by $\sqrt{(|2eEx/m|)}$.

2. An electron of mass m and charge $-e$ is emitted with negligible velocity from a thin straight wire, and thereafter moves under the action of a magnetic field of intensity B parallel to the wire and an electric field of potential Er, where t is the distance from the wire and E and B are constants. Show that the electron describes a cardioid with angular velocity $eB/2m$ about the wire,

3. A bead of mass m, and bearing a charge e, is free to slide on a smooth, circular, non-conducting loop of radius a, fixed horizontally. A vertical magnetic field has magnitude $B_0(t)a^2/(a^2+r^2)$, where t is the time, and r the radial distance in the horizontal plane from the centre of the wire. Show that if the bead is initially at rest, its velocity at time t, with an appropriate sign convention, is

$$v = \frac{ea}{2m}[B_0(0) - B_0(t)] \ln 2.$$

4. A magnetic field, parallel to the z-axis, whose magnitude depends only on distance from the z-axis, holds a charged particle in a circular orbit about the z-axis. Show that the speed of the particle can be increased by the action of the electric field induced by varying the magnetic field, without any change in the radius of the orbit, provided that the average magnetic field inside the orbit is twice the field at the orbit.

5. A radial electric field of uniform intensity $-E$ is maintained between a circular cylinder of radius a and a uniform wire lying along its axis, and a uniform magnetic field of intensity B is applied parallel to the wire. If electrons of mass m and charge $-e$ leave the wire with negligible speed, show that the path of such an electron is a cardioid and that none of the electrons will reach the cylinder if

$$B^2 > 8mE/(ea).$$

6. An ion of mass m and charge e moves in a gas in a constant electromagnetic field of vectors $\boldsymbol{E}$, $\boldsymbol{B}$ and undergoes a resistance of amount $\boldsymbol{v}/k$, where $\boldsymbol{v}$ is its velocity. Assuming a terminal velocity V for the motion of the ion, show that this is given by

$$\boldsymbol{V}(1+k^2e^2B^2) = ke\boldsymbol{E} + k^2e^2\boldsymbol{E}\times\boldsymbol{B} + k^3e^3\boldsymbol{B}(\boldsymbol{B}\cdot\boldsymbol{E}).$$

7. Current I flows along a circular cylinder of radius a, surrounded by a concentric conducting cylinder of radius b, which is maintained at a constant potential V above that of the inner cylinder. A uniform field B is applied parallel to the cylinders.

Electrons of mass m and charge $-e$ are released with negligible velocity from the inner cylinder. Neglecting variation of mass with velocity, show that, in cylindrical polar coordinates (r, θ, z)

$$\dot{\theta} = \frac{eB}{2m}(1-a^2/r^2), \quad \dot{z} = -\frac{\mu_0 Ie}{2\pi m}\ln r/a.$$

[P.T.O.

From energy considerations, or otherwise, show that if electrons are to reach the outer cylinder,

$$V \geqslant \frac{e}{2m}\left[\left\{\frac{B}{2b}(b^2-a^2)\right\}^2+\left\{\frac{\mu_0 I}{2\pi}\ln\left(\frac{b}{a}\right)\right\}^2\right].$$

8. A particle of mass m, carrying a charge q, moves in free space in an electric field $\mathbf{E} = -\textbf{grad}\, f(r)$ and in a uniform magnetic field of induction $B_r = B_\theta = 0$, $B_z = B$. (r, θ and z are cylindrical polar coordinates.) At time $t = 0$, $\dot{z} = 0$. Show that

$$z, \quad \tfrac{1}{2}m(\dot{r}^2+r^2\dot{\theta}^2)+qf(r), \quad mr^2\dot{\theta}+\tfrac{1}{2}qBr^2$$

remain constant throughout the motion.

An infinite earthed wire of negligible radius lies along the axis of an infinite circular cylinder of radius a whose potential is V; a uniform magnetic field of induction B is applied parallel to the wire. An electron of charge $-e(e > 0)$ and mass m is emitted from the wire with zero velocity. Show that it cannot reach the cylinder if

$$V < \frac{ea^2B^2}{8m}.$$

9. Prove that the path of a particle moving in a uniform magnetic field $B_0\mathbf{k}$ is a helix of the form

$$x = (U/\omega)\sin\omega t, \quad y = (U/\omega)\cos\omega t, \quad z = Wt,$$

where $\omega = eB_0/m$, and $\{U\,0\,W\}$ are the components of the velocity of projection. If the field is non-uniform, being given by

$$\mathbf{B} = B_0(\varepsilon y\mathbf{i}+\varepsilon x\mathbf{j}+\mathbf{k}),$$

where ε is small, prove that the components $\{u\ v\ w\}$ of the velocity at time t are approximately,

$$u = U\cos\omega t, \quad v = -U(1-\varepsilon W/\omega)\sin\omega t, \quad w = W,$$

the circumstances of projection being the same as above. (The unit vectors $\mathbf{i}, \mathbf{j}, \mathbf{k}$ denote the directions of the coordinate axes.)

10. A particle of mass m and charge e moves in the magnetic field of a current J flowing in an infinite straight wire.

Show that

(i) the angular momentum of the particle about the axis of the wire is constant,

(ii) $mv\cos\theta - \dfrac{e\mu_0 J}{2\pi}\ln r = \text{constant}$,

where v is the speed of the particle, r its distance from the axis of the wire, and θ the angle its direction of motion makes with the direction of the current.

Show further that if the particle be projected from a point P distant a from the wire with velocity u in the direction of the magnetic field at P, the trajectory will be confined to the region of space $a \leqslant r \leqslant b$, where b is the positive root of the equation

$$1-\frac{a^2}{r^2} = \left(\frac{e\mu_0 J}{2\pi mu}\right)^2\left(\ln\frac{r}{a}\right)^2$$

other than a.

14.3 Charged particles and currents

Since electric currents consist essentially of moving electric charges, some insight into certain problems of current flows can be obtained by considering the motion of a number of electric charges under the influence of electromagnetic fields. A complete discussion of the dynamics of a fully or partially ionized gas (or *plasma*) is beyond the scope of this book; as indicated on p. 569 the mutual interactions between the moving particles may profoundly influence the results which those for an isolated particle might suggest. However, we derive below two results of physical importance from the elementary equations.

1. The Poynting vector

Since the force acting on a charged particle is, with the usual notation, $\boldsymbol{F} = e(\boldsymbol{E}+\boldsymbol{v}\times\boldsymbol{B})$, the rate of working by the electromagnetic field on the particle is $\boldsymbol{F}\cdot\boldsymbol{v}$, i.e. $e\boldsymbol{E}\cdot\boldsymbol{v}$. Suppose now that in an element $d\tau$ of the volume contained within a closed surface T there are N charges of magnitude $+e$, the velocity of the rth of these charges being $\boldsymbol{V}_r$, and n charges of magnitude $-e$, the velocity of the sth of these being $\boldsymbol{v}_s$. Then the rate of working of the electromagnetic field on all the charges within the element $d\tau$ is

$$\boldsymbol{E}\cdot\left(e\sum_1^N \boldsymbol{V}_r - e\sum_1^n \boldsymbol{v}_s\right).$$

But the electric current $\boldsymbol{j}$ is defined by

$$e\sum_1^N \boldsymbol{V}_r - e\sum_1^n \boldsymbol{v}_s = \boldsymbol{j}\, d\tau$$

(see Vol. 1, p. 153). It follows that the rate of working of the electromagnetic field on the charges within $d\tau$ is $\boldsymbol{j}\cdot\boldsymbol{E}\, d\tau$ and so the rate of working on the charges within T is

$$\iiint_T \boldsymbol{j}\cdot\boldsymbol{E}\, d\tau.$$

Here, of course, $\boldsymbol{E}$ is the total electric field. Then it follows by the results of Vol. 2, pp. 425–427 that the Poynting vector $\boldsymbol{E}\times\boldsymbol{H}$ may be interpreted as the intensity of flow of electromagnetic energy.

2. Current limited by space charge

The emission of electrons and ions from charged conductors is important in various problems of electronics. We now consider the steady flow *in vacuo* of a current in which only electrons or ions are present so that the current is carried by charges of one sign only and the net space charge ϱ in the neighbourhood of charges is not zero. Further, we assume that the motion of the charges is so slow that the electrostatic potential V and field $\boldsymbol{E}$ can be calculated as if the charges were stationary. Then Poisson's equation must be satisfied so that

$$\nabla^2 V = -\varrho/\varepsilon_0. \tag{14.5}$$

If each particle has mass m and gains its energy solely from the electric field, then the energy equation of a typical particle is

$$\tfrac{1}{2}mv^2 = e(V_0 - V), \tag{14.6}$$

where V_0 is the electrostatic potential at the point of emission of the particle. Further, the current density $\boldsymbol{j}$ at any point is given by

$$\boldsymbol{j} = \varrho\boldsymbol{v}. \tag{14.7}$$

Eliminating ϱ and $\boldsymbol{v}$ from eqns. (14.5)–(14.7) gives a differential equation for V.

Unidirectional flow

As a special case we consider a one-dimensional flow. Suppose that the charges are emitted freely in unlimited quantity from the plane $x = 0$, at which $V = V_0$, and move parallel to Ox, the current per unit area flowing parallel to Ox being I. Then in this case the equation for V is

$$\frac{d^2V}{dx^2} = -\frac{I}{\varepsilon_0}\sqrt{\left\{\frac{m}{2e(V_0 - V)}\right\}}. \tag{14.8}$$

Since in the steady state charges are emitted at the plane $x = 0$ until the electric field vanishes there (i.e. until the field can no longer move the charges away from the plane) the boundary conditions are

$$V = V_0, \quad \frac{dV}{dx} = 0 \quad \text{at } x = 0.$$

The first integral of eqn. (14.8), obtained after multiplying by the integrating factor $\mathrm{d}V/\mathrm{d}x$, is

$$\left(\frac{\mathrm{d}V}{\mathrm{d}x}\right)^2 = \frac{4I}{\varepsilon_0}\sqrt{\left\{\frac{m(V_0-V)}{2e}\right\}}.$$

A further integration, after taking the square root, gives

$$(V_0-V)^{3/2} = \frac{9I}{4\varepsilon_0}\sqrt{\left(\frac{m}{2e}\right)}x^2. \tag{14.9}$$

If $V = 0$ when $x = a$, so that the plane $x = a$ is an earthed plate, eqn. (14.9) gives

$$I = \frac{4\varepsilon_0}{9}\left(\frac{2e}{m}\right)^{1/2}\frac{V_0^{3/2}}{a^2}. \tag{14.10}$$

Equation (14.10), known as *Child's equation*, shows that the current varies as the voltage to the power $\frac{3}{2}$. The effect of the space charge is to retard the flow of current which is therefore said to be *space-charge* limited. Note that since m is much larger for ions than for electrons the space-charge limitation effect is much greater for ions than electrons.

Example. Electrons of mass m and charge $-e$ are liberated with negligible velocity from an infinite cylindrical conductor of radius a at zero potential and pass towards an outer concentric cylindrical conductor maintained at a higher potential. Treating the particles in the space between the cylinders as a continuous distribution of negative charge, show that when a steady current I per unit length of the cylinder is flowing, the potential V at a distance r from the axes of the cylinders satisfies the equation

$$\frac{\mathrm{d}^2V}{\mathrm{d}x^2} = \frac{4V}{9y^2},$$

where

$$x = \ln(r/a), \qquad ry^2I = \frac{8\pi\varepsilon_0}{9}\left(\frac{2eV^3}{m}\right)^{1/2}.$$

Hence, show that y satisfies the differential equation

$$3y\frac{\mathrm{d}^2y}{\mathrm{d}x^2} = 1-y^2-\left(\frac{\mathrm{d}y}{\mathrm{d}x}\right)^2-4y\left(\frac{\mathrm{d}y}{\mathrm{d}x}\right)$$

and find the three coefficients A, B, C of the series solution

$$y = Ax+Bx^2+Cx^3+\ldots$$

This example is a much closer approximation to cases which occur in practice where electrons are emitted from a long cylindrical cathode (of small radius) and accelerated by a radial electric field so as to move towards a coaxial (larger) cylindrical anode.

Using cylindrical polar coordinates as specified, and assuming axial symmetry so that all variables depend on r only, eqns. (14.5)–(14.7) become

$$\frac{d^2V}{dr^2}+\frac{1}{r}\frac{dV}{dr}=-\frac{\varrho}{\varepsilon_0}, \tag{1}$$

$$\tfrac{1}{2}mv_r^2 = eV, \tag{2}$$

$$I=-2\pi r\varrho v_r, \tag{3}$$

where v_r is the (radial) velocity, I is the total current emitted per unit length of the cylinders. Elimination of ϱ, v_r from eqns. (1)–(3) gives

$$r\frac{d^2V}{dr^2}+\frac{dV}{dr}=\frac{I}{2\pi\varepsilon_0}\sqrt{\left\{\frac{m}{2eV}\right\}}. \tag{4}$$

Making the substitution $r = ae^x$ so that

$$r\,dV/dr = dV/dx, \qquad r^2\,d^2V/dr^2 = d^2V/dx^2 - dV/dx,$$

eqn. (4) becomes

$$\frac{d^2V}{dx^2}=\frac{rI}{2\pi\varepsilon_0}\sqrt{\left(\frac{m}{2eV}\right)}. \tag{5}$$

Then *defining* y by the relation

$$ry^2I=\frac{8\pi\varepsilon_0}{9}\left(\frac{2eV^3}{m}\right)^{1/2} \tag{6}$$

transforms eqn. (5) into

$$\frac{d^2V}{dx^2}=\frac{4V}{9y^2} \tag{7}$$

as required.

To obtain the equation for y, we eliminate V from eqn. (7) and eqn. (6), which can be written $V = ke^{2x/3}y^{4/3}$, where k is constant. We find

$$\frac{1}{V}\frac{d^2V}{dx^2}=\frac{4}{9}+\frac{16}{9y}\frac{dy}{dx}+\frac{4}{3y}\frac{d^2y}{dx^2}+\frac{4}{9y^2}\left(\frac{dy}{dx}\right)^2$$

and substitution in eqn. (7) gives

$$3y\frac{d^2y}{dx^2}=1-y^2-\left(\frac{dy}{dx}\right)^2-4y\frac{dy}{dx} \tag{8}$$

as required.

Assuming the series solution specified and equating coefficients in the usual manner, we find the solution of eqn. (8) to be

$$y = x-\frac{2}{5}x^2+\frac{11}{120}x^3-\frac{47}{3300}x^4+\dots$$

after using the initial conditions $y = 0$, $dy/dx = 1$ at $x = 0$. These conditions follow from the fact that $V = 0 = dV/dr$ when $r = a$, i.e. $V = 0 = dV/dx$ when $x = 0$.

14.4 Relativistic motion of charges

The correct equation of motion for a charge which may be moving at a speed comparable to that of light was obtained in Vol. 2, p. 359, in the form

$$\frac{d\boldsymbol{p}}{dt}=e(\boldsymbol{E}+\boldsymbol{v}\times\boldsymbol{B}), \tag{14.11}$$

where

$$\boldsymbol{p} = m\boldsymbol{v} = \frac{m_0\boldsymbol{v}}{\sqrt{(1-v^2/c^2)}}. \tag{14.12}$$

These equations of motion are far from easy to solve in general, and we shall confine ourselves in this section to cases where the electric and magnetic fields are constant both in time and space. The first step is to derive the change of the mass of the particle with time, that is, we calculate

$$\frac{\mathrm{d}m}{\mathrm{d}t} = \frac{\mathrm{d}}{\mathrm{d}t}\left\{\frac{m_0}{\sqrt{(1-v^2/c^2)}}\right\} = \frac{m_0\boldsymbol{v}\cdot\dot{\boldsymbol{v}}/c^2}{(1-v^2/c^2)^{3/2}}. \tag{14.13}$$

In order to do this it is only necessary to notice that

$$\mathrm{d}\boldsymbol{p} = \boldsymbol{v}\,\mathrm{d}m + m\,\mathrm{d}\boldsymbol{v} = \boldsymbol{v}\frac{m_0\boldsymbol{v}\cdot\mathrm{d}\boldsymbol{v}/c^2}{(1-v^2/c^2)^{3/2}} + \frac{m_0\,\mathrm{d}\boldsymbol{v}}{\sqrt{(1-v^2/c^2)}}.$$

Hence

$$\boldsymbol{v}\cdot\mathrm{d}\boldsymbol{p}/c^2 = \mathrm{d}m,$$

so that by using the original equation of motion the magnetic field drops out and the result

$$\frac{\mathrm{d}m}{\mathrm{d}t} = \frac{e\boldsymbol{E}\cdot\boldsymbol{v}}{c^2} \tag{14.14}$$

follows. Thus the rate of change of mass is determined only by the work done by the *electric* field on the particle. This is because the magnetic force acts on the particle in a direction at right angles to the direction of motion.

Consider now a charge in a uniform constant electric field $\boldsymbol{E}$, when the magnetic field is zero. The direction of the field may be chosen as one of the coordinate axes, say the x-axis. It is clear from the equation of motion that the path will lie in one plane which may be chosen as the plane Oxy. With these conventions the equation of motion can be written [if $\boldsymbol{p} = \{p_1\, p_2\, p_3\}$]

$$\dot{p}_1 = eE, \quad \dot{p}_2 = 0$$

and these may be integrated once to give

$$p_1 = eEt + c_1, \quad p_2 = c_2.$$

It is clear from the form of the first of these equations that, for some value of the time, the x-component of the momentum vanishes. It is therefore convenient to choose this moment as the zero for time reckoning and use the simpler solution in the form

$$p_1 = eEt, \quad p_2 = c_2. \tag{14.15}$$

But the mass of the particle may be written in the form

$$m = \sqrt{(m_0^2+p^2/c^2)}, \tag{14.16}$$

and so, by inserting this into eqn. (14.15) the mass is determined as a function of the time in the form

$$m = \sqrt{\{m_0^2+(c_2^2/c)^2+(eEt)^2/c^2\}} = \{\bar{m}_0^2+(eEt)^2/c^2\}, \tag{14.17}$$

where $\bar{m}_0$ denotes the mass at time $t = 0$. Now the velocity of the particle has the form $\boldsymbol{v} = \boldsymbol{p}/m$, and, since this is the time derivative of the position, the x-coordinate can be found by one integration thus:

$$\frac{\mathrm{d}x}{\mathrm{d}t} = \frac{eEt}{\sqrt{\{\bar{m}_0^2+(eEt)^2/c^2\}}}, \quad x = \frac{c^2}{eE}\sqrt{\{\bar{m}_0^2+(eEt)^2/c^2\}}. \tag{14.18}$$

Here the origin has been chosen as the initial position of the particle so that there is no constant of integration. In the same way y is easily found, thus:

$$\frac{\mathrm{d}y}{\mathrm{d}t} = \frac{c_2}{\sqrt{\{\bar{m}_0^2+(eEt)^2/c^2\}}}, \quad y = \frac{cc_2}{eE}\sinh^{-1}\left(\frac{eEt}{\bar{m}_0 c}\right).$$

By eliminating t the equation of the path is found to have the form

$$x = \frac{c^2 m_0}{eE}\cosh\left(\frac{eEy}{cc_2}\right), \tag{14.19}$$

that is, a catenary.

Consider next the case when there is a magnetic field but no electric field. Choose the direction of the field as the z-axis. The equation of motion now takes the form

$$\dot{\boldsymbol{p}} = \mu_0 e\boldsymbol{v}\times\boldsymbol{H}.$$

However, m is now constant (so that $|\boldsymbol{v}|$ must be constant) and so

$$\frac{\mathrm{d}\boldsymbol{v}}{\mathrm{d}t} = \frac{\mu_0 e}{m}\boldsymbol{v}\times\boldsymbol{H},$$

i.e. $$\dot{v}_x = \frac{\mu_0 eH}{m}v_y, \quad \dot{v}_y = -\frac{\mu_0 eH}{m}v_x, \quad \dot{v}_z = 0.$$

These are just the equations in the non-relativistic case, except that m has a different (but constant) value. We integrate them in the same way. If the second of these equations is multiplied by i and added to the first, we obtain

$$\frac{\mathrm{d}}{\mathrm{d}t}(v_x+\mathrm{i}v_y) = -\mathrm{i}\omega(v_x+\mathrm{i}v_y), \quad \text{where} \quad \omega = \frac{\mu_0 eH}{m},$$

which can be integrated at once. Writing the solution in the form

$$v_x + iv_y = ae^{-i\omega t}, \quad a = be^{-i\alpha},$$

where b, α are real, it follows that

$$v_x = b\cos(\omega t+\alpha), \quad v_y = -b\sin(\omega t+\alpha). \tag{14.20}$$

The results which have just been found can easily be integrated again and show that the projection of the motion on the $x-y$ plane is a circle. At the same time, from the third equation of motion it follows that

$$z = z_0 + kt,$$

corresponding to a uniform motion up the z-axis. The total path is therefore again that of a circular helix.

It remains to consider the case when both electric and magnetic fields are present. The interesting special case here is that in which the fields are perpendicular. Even without this restriction one way of dealing with the problem is by successive approximation. We first of all solve the non-relativistic equation as in § 14.2 above, which is straightforward because the mass is constant. The Newtonian solution may then be used as an approximate solution for determining the variation of the mass and the resulting equations integrated again. This is not, however, a very satisfactory way of proceeding, for it requires that the velocity of the particle shall remain very small and this is found to require the component of the electric field perpendicular to the magnetic field to be very small. In the case in which we are most interested the electric field is entirely perpendicular to the magnetic field.

A better way of proceeding is to use the transformation of electric and magnetic fields found in Chapter 12. It was shown there that any pair of mutually perpendicular electric and magnetic fields could be transformed into either a purely electric or a purely magnetic field by a Lorentz transformation. The only restriction is that if the fields are mutually perpendicular and a certain relation exists between their magnitudes (i.e. $\mu_0 \boldsymbol{H}^2 = \varepsilon_0 \boldsymbol{E}^2$) this transformation is impossible. For perpendicular fields, then, it is only necessary to consider this limiting case. This can be done as follows: choosing the coordinate axes so that the $\boldsymbol{E}$-field is parallel to the y-axis, and the $\boldsymbol{H}$-field is parallel to the z-axis, we have

$$\begin{aligned} \dot{p}_x &= e\mu_0 v_y H = \frac{ev_y}{c}E, \\ \dot{p}_y &= eE - e\mu_0 v_x H = eE\left(1-\frac{v_x}{c}\right), \\ \dot{p}_z &= 0. \end{aligned} \tag{14.21}$$

(Here we have used the fact that $H = \sqrt{(\varepsilon_0/\mu_0)}E$, and also $c^2\mu_0\varepsilon_0 = 1$.)

These equations have the usual consequence that

$$\frac{dm}{dt} = \frac{e}{c^2}\boldsymbol{E}\cdot\boldsymbol{v} = \frac{e}{c^2}Ev_y.$$

Accordingly, on integrating,

$$\begin{aligned} p_z &= \text{constant}, \\ p_x - mc &= \text{constant} = -\alpha \text{ (say).} \end{aligned} \tag{14.22}$$

Moreover,

$$m^2c^2 = m_0^2c^2 + \boldsymbol{p}^2,$$

so that

$$m^2c^2 - p_x^2 = m_0^2c^2 + p_y^2 + p_z^2 = p_y^2 + k^2 \quad \text{(say)}, \tag{14.23}$$

where k is constant. Hence

$$(mc+p_x)\alpha = p_y^2 + k^2, \tag{14.24}$$

so that

$$mc + p_x = \frac{p_y^2 + k^2}{\alpha},$$

whilst

$$mc - p_x = \alpha$$

as before. Hence

$$\begin{aligned} mc &= \frac{1}{2}\left(\alpha + \frac{p_y^2+k^2}{\alpha}\right), \\ p_x &= \frac{1}{2}\left(\frac{p_y^2+k^2}{\alpha} - \alpha\right). \end{aligned} \tag{14.25}$$

Moreover, since $p_x = mv_x$, the second equation of motion becomes

$$m\dot{p}_y = eE\left(m - \frac{p_x}{c}\right) = \frac{eE\alpha}{c}. \tag{14.26}$$

This now gives a differential equation that will connect p_y and t. For since $mc = p_x + \alpha$, and p_x is given in terms of p_y, it follows that

$$\left\{\alpha + \frac{1}{2}\,\frac{p_y^2+k^2}{\alpha} - \frac{1}{2}\alpha\right\}\dot{p}_y = eE\alpha, \tag{14.27}$$

so

$$\alpha p_y + \frac{1}{3}\,\frac{p_y^3}{\alpha} + \frac{k^2}{\alpha}p_y = 2eE\alpha t. \tag{14.28}$$

(Here we are assuming that $p_y = 0$ when $t = 0$.) The rest of the determination of the trajectory is performed by changing the variable from t to p_y, for

$$m\,\mathrm{d}p_y = \frac{eE\alpha}{c}\,\mathrm{d}t,$$

and

$$m\frac{\mathrm{d}x}{\mathrm{d}t} = p_x,$$

so that

$$\frac{\mathrm{d}x}{\mathrm{d}p_y} = \frac{cp_x}{eE\alpha} = \frac{1}{2}\frac{c}{eE\alpha}\left\{\frac{p_y^2+k^2}{\alpha}-\alpha\right\}.$$

Hence

$$x = \frac{c}{2eE\alpha^2}(k^2-\alpha^2)p_y + \frac{cp_y^3}{6eE\alpha^2}, \tag{14.29}$$

and similarly,

$$y = \frac{c^2}{2eE\alpha}p_y^2, \quad z = \frac{p_z c^2}{eE\alpha}p_y, \tag{14.30}$$

which define a twisted cubic curve in parametric form.

Example. A particle of rest mass m and charge e moves in a uniform magnetic field of flux density $\boldsymbol{B}$. Show that the velocity $\boldsymbol{v}$ of the particle has a constant magnitude v and that the particle describes a helix with axis parallel to the magnetic field. If the angle between $\boldsymbol{v}$ and $\boldsymbol{B}$ is α, show that the helix lies on a cylinder of radius

$$(mcv \sin\alpha)/[eB\{1-(v/c)^2\}^{1/2}].$$

By means of a moving frame of reference, or otherwise, show that if such a particle moves in uniform fields of intensities E and B, at right angles to each other and such that $cB > E$, and the particle starts from rest at $t = 0$, its mass at time t is

$$\frac{m\{B^2-(E^2/c^2)\cos\theta\}}{B^2-(E^2/c^2)}$$

where

$$te\{B^2-(E^2/c^2)\}^{3/2} = m\{B^2\theta-(E^2/c^2)\sin\theta\}.$$

Let $M = \dfrac{m}{\sqrt{(1-v^2/c^2)}}$ be the relative mass. The equation of motion is

$$\frac{\mathrm{d}}{\mathrm{d}t}(M\dot{\boldsymbol{r}}) = e\dot{\boldsymbol{r}}\times\boldsymbol{B}.$$

Hence

$$M\dot{\boldsymbol{r}}\cdot\frac{\mathrm{d}}{\mathrm{d}t}(M\dot{\boldsymbol{r}}) = 0$$

so

$$M\dot{\boldsymbol{r}}^2 = \text{constant},$$

i.e.

$$\frac{m\dot{\boldsymbol{r}}^2}{1-\dot{\boldsymbol{r}}^2/c^2} = \text{constant},$$

which implies $\dot{\boldsymbol{r}}^2 = \text{constant}$ and so $M = \text{constant}$.

We return to the original equation and integrate once:

$$M\dot{\boldsymbol{r}} = e\boldsymbol{r}\times\boldsymbol{B}+\boldsymbol{c},$$

so that

$$\boldsymbol{B}\cdot\dot{\boldsymbol{r}} = \boldsymbol{c}\cdot\boldsymbol{B}/M = \text{constant}.$$

Since $\dot{\boldsymbol{r}}^2$ is also constant, this proves that the tangent to the path makes a constant angle with the fixed direction $\boldsymbol{B}$, i.e. that the path is a helix.

Taking $\boldsymbol{B} = B\boldsymbol{k}$, by a suitable choice of axes, we then have

$$M\dot{x} = eBy+c_1,$$
$$M\dot{y} = -eBx+c_2,$$
$$M\dot{z} = c_3,$$

so that the z-component of velocity is also constant. By choosing the origin suitably, it is obviously possible to make c_1 and c_2 vanish. Then

$$M^2\ddot{x} = -e^2B^2x,$$

so that

$$x = P\cos\left(\frac{eBt}{M}\right)+Q\sin\left(\frac{eBt}{M}\right) = R\cos\left(\frac{eBt}{M}-\phi\right), \quad \text{(say)} \tag{1}$$

with

$$y = M\dot{x}/(eB) = -R\sin\left(\frac{eBt}{M}-\phi\right), \tag{2}$$

In order to determine R, we observe that

$$\dot{x}^2+\dot{y}^2 = \frac{R^2e^2B^2}{M^2} = v^2\sin^2\alpha.$$

$$\therefore \quad R = \frac{Mv\sin\alpha}{eB} = \frac{mv\sin\alpha}{eB\sqrt{(1-v^2/c^2)}}. \tag{3}$$

When E and B are at right angles and $cB > E$ it is possible to choose a frame of reference in which the electric field vanishes, for if $\boldsymbol{B} = B\boldsymbol{k}$, $\boldsymbol{E} = E\boldsymbol{j}$ and we perform a Lorentz transformation along the x-axis, with velocity V, then in general

$$E_1' = E_1, \qquad B_1' = B_1,$$
$$E_2' = \beta(E_2-VB_3), \qquad B_2' = \beta(B_2+VE_3/c^2),$$
$$E_3' = \beta(E_3+VB_2) \qquad B_3' = \beta(B_3-VE_2/c^2),$$

which give, with the special choice $V = E/B$,

$$\boldsymbol{E}' = 0, \quad \boldsymbol{B}' = \{0 \;\; 0 \;\; B/\beta\}.$$

In the primed coordinate system the initial velocity of the particle is

$$\{-E/B \;\; 0 \;\; 0\}$$

so that

$$v'^2 = E^2/B^2.$$

In this frame $O'x'y'z'$ there is no electric field and so the motion corresponds to tha discussed above leading to eqns. (1), (2) and (3). Since the initial velocity of the particle in this frame, is $\{-E/B \;\; 0 \;\; 0\}$ the phase constant, ϕ, in eqn. (1) is π, and the (constant speed of the particle is $v' = E/B$. Because $\dot{z}' = 0$ initially the motion takes place entirely in the plane $z' = 0$, i.e. $\alpha = \pi/2$, and is the circle

$$x' = -R' \sin\left(\frac{eB't'}{M'}\right), \quad y' = R' \cos\left(\frac{eB't'}{M'}\right), \quad z' = 0 \tag{4}$$

corresponding to (1) and (2). The radius of this circle is

$$R' = \frac{M'v'}{eB'} = \frac{mv'}{eB'(1-v'^2/c^2)^{1/2}} = \frac{mE}{eBB'\{1-E^2/(B^2c^2)\}^{1/2}}.$$

Since
$$B' = \beta\{B - E^2/(Bc^2)\}^{1/2} = (B^2 - E^2/c^2)^{1/2},$$

$$R' = \frac{mE}{e(B^2-E^2/c^2)^{1/2}}.$$

The information concerning the motion is required in terms of observations made in the frame $Oxyz$ and t.

$$\therefore \quad t = \beta(t' + Vx'/c^2) = \left\{t' - \frac{mE^2}{eB(B^2c^2-E^2)} \sin\left(\frac{eB't'}{M'}\right)\right\}\{1-E^2/(B^2c^2)\}^{-1/2}$$

We put
$$\theta = \frac{eB't'}{M'} = \frac{e}{mB}\left(B^2 - \frac{E^2}{c^2}\right)t',$$

so that
$$t' = \frac{mB\theta}{e(B^2-E^2/c^2)},$$

and
$$t = \frac{m\{B^2\theta - (E^2/c^2)\sin\theta\}}{e(B^2-E^2/c^2)^{3/2}}.$$

In order to calculate the mass we must know the velocity; and we can easily find this in the frame $O'x'y'z'$ and transform it to the frame $Oxyz$ by using the addition formulae for velocity (see Vol. 2, § 9.2).

$$v_x = \frac{v_x' + E/B}{1 + v_x'E/(Bc^2)}, \quad v_y = \frac{v_y'\{1-E^2/(B^2c^2)\}^{1/2}}{1+v_x'E/(Bc^2)}.$$

Now
$$v_x' = \frac{dx'}{dt'} = -\frac{R'eB'}{M'}\cos\theta = -\frac{E}{B}\cos\theta,$$

$$v_y' = \frac{dy'}{dt'} = -\frac{R'eB'}{M'}\sin\theta = -\frac{E}{B}\sin\theta.$$

$$\therefore \quad v_x = \frac{(E/B)(1-\cos\theta)}{1-(E^2/B^2c^2)\cos\theta}, \quad v_y = -\frac{(E/B)\sin\theta\{1-E^2/(B^2c^2)\}^{1/2}}{1-(E^2/B^2c^2)\cos\theta}$$

giving
$$v^2 = v_x^2 + v_y^2$$
$$= \frac{E^2B^2}{\{B^2-(E^2/c^2)\cos\theta\}^2}\left\{(1-\cos\theta)^2 + \left(1-\frac{E^2}{B^2c^2}\right)\sin^2\theta\right\}.$$

Hence
$$1 - \frac{v^2}{c^2} = \frac{(B^2-E^2/c^2)^2}{\{B^2-(E^2/c^2)\cos\theta\}^2}$$

and
$$M = \frac{m}{(1-v^2/c^2)^{1/2}} = \frac{m\{B^2-(E^2/c^2)\cos\theta\}}{B^2-E^2/c^2}.$$

Miscellaneous Exercises XIV

1. Solve the problem of a particle moving relativistically in an electric field $E\boldsymbol{j}$ and a magnetic field $B\boldsymbol{i}$ by successive approximation.

Find conditions for your solution to be valid (so that $v \ll c$ throughout the motion).

Contrast the first approximation found in the case when $\mu_0 H^2 = \varepsilon_0 E^2$ with the solution found in the text.

2. A stream of electrons, each of mass m and charge $-e$, is emitted normally from a cathode lying in the yz-plane, and moves under the action of an electric field of variable intensity $E(x)$, parallel to the x-axis, and a magnetic field of uniform intensity $m\omega/e$, parallel to the z-axis, ω being a constant. If u, v are the x, y components of the velocity of the electrons at any point, show that in the steady state

$$u\frac{du}{dx} = -\frac{e}{m}E(x) - \omega v, \quad \text{and} \quad \frac{dv}{dx} = \omega, \quad u \neq 0.$$

3. An electron of mass m, charge $-e$ and velocity $\boldsymbol{v}$, moving in a constant uniform magnetic field $B\boldsymbol{k}$, experiences a force

$$-eB\boldsymbol{v}\times\boldsymbol{k},$$

where $\boldsymbol{k}$ is a fixed unit vector. Prove that if the electron is subject also to a force $\boldsymbol{F} = f(r)\boldsymbol{r}$, where $\boldsymbol{r}$ is the position vector of the electron relative to a fixed origin O, the equation of its motion relative to a frame rotating with constant angular velocity $\omega\boldsymbol{k}$ about O is

$$m\{\ddot{\boldsymbol{r}} + 2\omega\boldsymbol{k}\times\dot{\boldsymbol{r}} + \omega^2\boldsymbol{k}\times(\boldsymbol{k}\times\boldsymbol{r})\} = \boldsymbol{F} - eB\dot{\boldsymbol{r}}\times\boldsymbol{k} - eB\omega(\boldsymbol{k}\times\boldsymbol{r})\times\boldsymbol{k}.$$

Hence show that if eB/m is small the effect of the magnetic field is, to the first order in eB/m, to cause electron orbits to precess with angular velocity $(eB/2m)\boldsymbol{k}$.

When there is no magnetic field an electron moves in simple harmonic motion with frequency n along a straight line l. Show that in a constant uniform magnetic field, inclined at an acute angle α to l and of magnitude B such that eB/m is small, the motion of the electron is composed of vibrations with frequencies n, $n+n'$ and $n-n'$, where

$$n' = eB/(4\pi m).$$

4. Electrons leave the surface of a conducting circular cylinder of radius a with negligible initial velocity and are accelerated towards the surface of a coaxial conducting cylinder of radius b ($b > a$) by means of a radial electric field. The difference in potential between the cylinders is V and a current I flows along the inner cylinder. Neglecting effects due to space charge, show that if a uniform magnetic field B is applied parallel to the axis of the cylinders the electrons will fail to reach the outer cylinder if

$$V < \frac{e}{8m}\left\{\frac{\mu_0^2 I^2}{\pi^2}\left(\ln\frac{b}{a}\right)^2 + \frac{B^2}{b^2}(b^2-a^2)^2\right\}.$$

Show also that the radial component of velocity of an electron at distance r from the axis is given by

$$v^2 = \frac{2eV}{m\ln(b/a)}\ln(r/a) - \frac{e^2B^2}{4m^2r^2}(r^2-a^2)^2 - \frac{\mu_0^2 e^2 I^2}{4\pi^2 m^2}\{\ln(r/a)\}^2.$$

5. Assuming that the force on a unit charge moving with velocity $\boldsymbol{v}$ *in vacuo* in an electric field $\boldsymbol{E}$ and magnetic field of induction $\boldsymbol{B}$ is

$$\boldsymbol{E} + \boldsymbol{v}\times\boldsymbol{B},$$

show that the rate at which work is done by the field on a continuous distribution of charge moving in a region V bounded by a surface Σ is

$$-\frac{1}{\mu_0}\oint\!\!\oint_{\Sigma}(\boldsymbol{E}\times\boldsymbol{B})\cdot\mathrm{d}\boldsymbol{S}-\frac{1}{2}\frac{\mathrm{d}}{\mathrm{d}t}\iiint_V\left(\varepsilon_0\boldsymbol{E}^2+\frac{\boldsymbol{B}^2}{\mu_0}\right)\mathrm{d}\tau.$$

6. Electrons of charge e and mass m leave an infinite plane cathode $x = 0$ at zero potential with initial speed v_0 and zero initial acceleration, to impinge upon a parallel plane anode $x = a$ at potential V. Show that, if there is a steady thermionic current I across a unit area perpendicular to the x-direction, then the speed v of the electrons at distance x from the cathode is given by

$$(v-v_0)(v+2v_0)^2 = 9Iex^2/(2\varepsilon_0 m).$$

Hence, if v_0 is small in comparison with the speed of impact upon the anode, show that

$$I = \frac{4\varepsilon_0}{9a^2}\left(\frac{2e}{m}\right)^{1/2}V^{3/2}+\frac{8\varepsilon_0 v_0 V}{9a^2}.$$

7. A current I flows in a circular loop of radius a. Prove that a particle of mass m and charge q can move in a circular orbit of radius r ($<a$), concentric and coplanar with the loop, provided its angular velocity is

$$-\frac{\mu_0 Iq}{2ma}\sum_{n=0}^{\infty}\frac{(2n)!\,(2n+1)!}{2^{4n}(n!)^4}\left(\frac{r}{a}\right)^{2n}.$$

[The formulae

$$P_{2n}(0) = (-1)^n\frac{(2n)!}{2^{2n}(n!)^2},\qquad P'_{2n+1}(0) = (-1)^n\frac{(2n+1)!}{2^{2n}(n!)^2}$$

may be used without proof if required.]

8. A particle of mass m and electric charge q moves in the magnetic field produced by a constant current J in an infinitely long straight fixed wire. Derive the equations of motion of the particle in cylindrical polar coordinates r, θ, z with the wire as axis.

The particle is projected from a point at distance a from the wire with a velocity of magnitude v in the direction coplanar with and perpendicular to the wire and away from it. Show that in the subsequent motion r will vary between the limits $ae^{\pm k}$ and that the radius of curvature of the path is kr, where $k = \dfrac{2\pi mv}{q\mu_0 J}$. Show that when the particle is for the first time again moving directly away from the wire it will be displaced from its original position by a distance

$$4ka\int_0^1 \sinh\left(k\sqrt{(1-\lambda^2)}\right)\mathrm{d}\lambda$$

measured parallel to the wire.

9. A polar molecule, free to rotate in space about its centre under the influence of an electric field, is represented by two particles each of mass m, carrying equal and opposite charges $\pm e$, at the ends of a light rigid rod of length $2a$, whose mid-point is fixed.

The particle with charge $+e$ is assumed to be at

$$\{a\sin\theta\cos\psi\quad a\sin\theta\sin\psi\quad a\cos\theta\}$$

and that with charge $-e$ at

$$\{-a \sin \theta \cos \psi \quad -a \sin \theta \sin \psi \quad -a \cos \theta\}.$$

The electric field is taken to be of strength E in the z-direction. Show that the Lagrangian function for the system is

$$L = ma^2(\dot{\theta}^2+\dot{\psi}^2 \sin^2 \theta)+2eEa \cos \theta.$$

Obtain the Lagrangian equations of motion and prove that a steady motion is possible with $\theta = \pi/4$ if

$$\dot{\psi}^2 = \sqrt{2}eE/(ma).$$

ANSWERS TO THE EXERCISES

Exercises 11.2 (p. 454)

1. $\boldsymbol{E} = \omega\{\sin(\omega t+kz) \quad -\cos(\omega t+kz) \quad 0\}$

$\boldsymbol{B} = k\{-\cos(\omega t+kz) -\sin(\omega t+kz) \quad 0\}$

$\boldsymbol{E}' = \omega\{-\sin(\omega t-kz) -\cos(\omega t-kz) \quad 0\}$

$\boldsymbol{B}' = k\{-\cos(\omega t-kz) -\sin(\omega t-kz) \quad 0\}$

$|\boldsymbol{E}+\boldsymbol{E}'| = 2\omega \sin kz, \quad |\boldsymbol{B}+\boldsymbol{B}'| = 2k\cos kz.$

Exercises 11.6 (p. 472)

1. $v_1 = (\mu_1\varepsilon_1)^{-1/2}, \quad \dfrac{E_0'}{E_0} = \left|\dfrac{\mu_2-\mu_1 k v_1}{\mu_2+\mu_1 k v_1}\right|.$
2. $k^2 = \omega^2\mu\varepsilon - \mathrm{i}\omega\sigma\mu, \quad \boldsymbol{B} = (k/\omega)\,\boldsymbol{j}\exp\{\mathrm{i}(\omega t-kz)\}.$
3. Reflected wave:

$$E_y' = \mathscr{E}\,\frac{[\mu^2-(m^2+n^2)\,\mu_0^2]\cos\omega(t+x/c) - 2m\mu\mu_0\sin\omega(t+x/c)}{(\mu+n\mu_0)^2+m^2\mu_0^2} = -cB_z'.$$

Transmitted wave:

$$E_y'' = 2\mathscr{E}\,\mathrm{e}^{-m\omega x/c}\,\frac{(\mu+n\mu_0)\cos\omega(t-nx/c)+m\mu_0\sin\omega(t-nx/c)}{(\mu+n\mu_0)^2+m^2\mu_0^2},$$

$$B_z'' = 2\mathscr{E}\,\mathrm{e}^{-m\omega x/c}\,\frac{[n\mu+(n^2-m^2)\,\mu_0]\cos\omega(t-nx/c)+(m\mu+2mn\mu_0)\sin\omega(t-nx/c)}{(\mu+n\mu_0)^2+m^2\mu_0^2}.$$

4. $2\mathrm{e}^{-2\alpha\omega z/c}/(n+1).$

Exercises 11.7 (p. 490)

1. $\psi_0(z) = A\sin(\pi z/a), \quad \varkappa^2 = \nu^2/c^2 - \pi^2/a^2.$
2. See eqn. (11.136) with $A_{rs} = A, \quad \mu_{rs} = \beta, \quad \nu_{rs}^2 = \pi^2\left(\dfrac{m^2}{a^2}+\dfrac{n^2}{b^2}\right) = \nu^2.$
3. Energy-flow $\quad \dfrac{\beta\omega}{2\mu_0 c^2\nu^4}\left\{\dfrac{m^2\pi^2}{a^2}\cos^2\left(\dfrac{m\pi x}{a}\right)\sin^2\left(\dfrac{n\pi y}{b}\right)+\dfrac{n^2\pi^2}{b^2}\sin^2\left(\dfrac{m\pi x}{a}\right)\cos^2\left(\dfrac{n\pi y}{b}\right)\right\}.$
4. $\beta^2 = \omega^2/v^2 - n^2\pi^2/b^2, \quad \mu\omega:\ \beta:\ -in\pi/b.$
5. $X = \sin(r\pi x/a), \quad Y = \sin(s\pi y/b), \quad \beta^2 = \mu\varepsilon\omega^2 - \pi^2\left(\dfrac{r^2}{a^2}+\dfrac{s^2}{b^2}\right) > 0.$

Miscellaneous Exercises XI (p. 507)

5. $\alpha = r\pi/a, \quad r = 1, 2, 3, \ldots; \quad \omega > r\pi c/a.$

6. $\boldsymbol{i} = \{H_y \quad 0 \quad 0\}.$

7. $\boldsymbol{i} = \boldsymbol{k}(cA/a)\cos(\omega(t-z/c), \quad \sigma = (A/a)\cos\omega(t-z/c).$

8. $\beta = \varepsilon_0. \quad S = A\sin(r\pi x/a)\sin(s\pi y/b)\begin{bmatrix}\cos\\ \sin\end{bmatrix}\omega t.$

9. $k^2 = \omega^2/c^2 - \varkappa_r^2,$ where $J_0'(\varkappa_r a) = 0,$ i.e. $\varkappa_r a$ are the zeros of $J_1(\varkappa)$.

Miscellaneous Exercises XII (p. 534)

1(a). It can move in a direction perpendicular to $\boldsymbol{E}$ with a speed $v = |\boldsymbol{E}|/B_\perp$, where $B_\perp$ is the component of $\boldsymbol{B}$ in a direction at right angles to $\boldsymbol{v}$ in the plane containing $\boldsymbol{v}$ and $\boldsymbol{B}$.

Miscellaneous Exercises XIII (p. 566)

5. $\phi = \frac{1}{r}M\left(t-\frac{r}{c}\right).$

6. $E_x = \frac{1}{\mu_0}\frac{\partial^2 S}{\partial z\,\partial x}, \quad E_y = \frac{1}{\mu_0}\frac{\partial^2 S}{\partial y\,\partial z}, \quad E_z = \frac{1}{\mu_0}\left(\frac{\partial^2 S}{\partial z^2} - \frac{1}{c^2}\frac{\partial^2 S}{\partial t^2}\right).$

8. $V = (\boldsymbol{k}\cdot\boldsymbol{r})\left\{\frac{c}{r^3}f\left(t-\frac{r}{c}\right) + \frac{1}{r^2}\frac{\partial}{\partial t}f\left(t-\frac{r}{c}\right)\right\}.$

9. $\nabla^2\boldsymbol{A} - \mu K\varepsilon_0\frac{\partial^2\boldsymbol{A}}{\partial t^2} = -\mu\boldsymbol{j}, \quad \nabla^2\phi - \mu K\varepsilon_0\frac{\partial^2\phi}{\partial t^2} = 0.$

Miscellaneous Exercises XIV (p. 596)

9. $a\ddot{\theta} - a\dot{\varphi}^2\sin\theta\cos\theta + (eE/m)\sin\theta = 0, \quad \frac{d}{dt}(\dot{\varphi}\sin^2\theta) = 0.$

INDEX